T0298697

A Laboratory Course
in Nanoscience
and Nanotechnology

A Laboratory Course in Nanoscience and Nanotechnology

Dr. Gérrard Eddy Jai Poinern

Murdoch University
Perth, Western Australia

CRC Press
Taylor & Francis Group
Boca Raton   London   New York

CRC Press is an imprint of the
Taylor & Francis Group, an **Informa** business

CRC Press
Taylor & Francis Group
6000 Broken Sound Parkway NW, Suite 300
Boca Raton, FL 33487-2742

Printed on acid-free paper
Version Date: 20140707

International Standard Book Number-13: 978-1-4822-3103-8 (Hardback)

Library of Congress Cataloging-in-Publication Data

Poinern, Gerrard Eddy Jai.
   A laboratory course in nanoscience and nanotechnology / Gerrard Eddy Jai Poinern.
      pages cm
   Summary: "Written by a well-regarded teacher, this first-ever nanolab manual is aimed at the fast-growing number of practical nanoscience and nanotechnology modules springing up around the world. It is not a direct competitor for theoretical texts, though it does describe laboratory experiments in detail and link them to the theory to place them in context.
   This highly interdisciplinary book places a strong emphasis on developing skills to prepare undergraduate and graduate students from physics, chemistry, engineering, and biology backgrounds for work in nanoscience- and nanotechnology-related industries"-- Provided by publisher.
      Includes bibliographical references and index.
      ISBN 978-1-4822-3103-8 (hardback)
      1. Nanotechnology--Textbooks. 2. Nanotechnology--Laboratory manuals. I. Title.

T174.7.P65 2014
620'.5--dc23   2014024492

Visit the Taylor & Francis Web site at
http://www.taylorandfrancis.com

and the CRC Press Web site at
http://www.crcpress.com

For my wife, Varny; sons Nathaniell and Rouben;
and in memory of my parents, Antoine and Eileen

Contents

Foreword, xix

Preface, xxiii

About the Author, xxvii

List of Abbreviations, xxix

CHAPTER 1 ▪ The Nano World   1

  1.1   INTRODUCTION   1

  1.2   NANOTECHNOLOGY   2

  1.3   NANOSCIENCE   5

  1.4   NATURE'S BIOLOGICAL PATHWAY   6

  1.5   EXAMPLES OF NANOMATERIALS AND NANOSTRUCTURES FOUND IN NATURE   7

      1.5.1   The Beak of the Humbolt Squid (*Dosidicus gigas*)   7

      1.5.2   The Beard of the Mussel (*Mytilus edulis*)   8

      1.5.3   Superhydrophobic Properties Found in Some Plant Leaves   10

  1.6   OVERVIEW OF CHAPTERS   12

  REFERENCES   13

CHAPTER 2 ▪ Nanomaterials and Their Synthesis   15

  2.1   INTRODUCTION   15

  2.2   NANOMETER-SCALE MATERIALS   16

2.3   TYPES OF NANOMETER-SCALE MATERIALS   18

    2.3.1   Nanometer-Scale Metals   19

        2.3.1.1   Nanogold   20

        2.3.1.2   Nanosilver   21

        2.3.1.3   Nanocopper   22

        2.3.1.4   Nanoiron   22

    2.3.2   Nano Metal Oxides   23

        2.3.2.1   Aluminum Oxide   23

        2.3.2.2   Titanium Dioxide   24

        2.3.2.3   Zinc Oxide   24

        2.3.2.4   Iron Oxides   24

    2.3.3   Nanopolymers   25

    2.3.4   Quantum Dots   26

    2.3.5   Nanocarbons   27

        2.3.5.1   Carbon Nanotubes   28

        2.3.5.2   Graphene   30

2.4   SYNTHESIS OF NANOMETER-SCALE MATERIALS   31

    2.4.1   Introduction   31

    2.4.2   Top-Down Techniques   33

        2.4.2.1   Photolithography   33

        2.4.2.2   Molecular Beam Epitaxy   34

    2.4.3   Bottom-Up Techniques   34

        2.4.3.1   Colloids   34

        2.4.3.2   Sol-Gel   35

        2.4.3.3   Vapor Deposition and Chemical Vapor
                 Deposition   36

        2.4.3.4   Sputtering   36

        2.4.3.5   Laser Ablation   36

        2.4.3.6   Anodic Aluminum Oxide Templates   37

        2.4.3.7   Spray Pyrolysis   38

        2.4.3.8   Ultrasonic Synthesis   38

        2.4.3.9   Microwave Synthesis   39

REFERENCES   39

CHAPTER 3 ■ Characterization Methods for Studying Nanomaterials 41

3.1 INTRODUCTION 41
3.2 SCANNING ELECTRON MICROSCOPY 43
3.3 TRANSMISSION ELECTRON MICROSCOPY 46
3.4 SCANNING TUNNELING MICROSCOPY 47
3.5 ATOMIC FORCE MICROSCOPY 49
3.6 X-RAY DIFFRACTION 53
3.7 UV-VIS SPECTROSCOPY 57
3.8 THIN-LAYER CHROMATOGRAPHY 58
3.9 RAMAN SPECTROSCOPY 60
3.10 DYNAMIC LIGHT SCATTERING 61
REFERENCES 63

CHAPTER 4 ■ Laboratory Safety and Scientific Report Writing 65

4.1 CHAPTER OVERVIEW 65
4.2 INTRODUCTION TO LABORATORY SAFETY 65
4.2.1 Good Laboratory Practices 66
4.2.2 Preparation 66
4.2.3 Protective Clothing 69
4.2.4 Eye Protection 69
4.2.5 Laboratory Hazards 70
4.2.5.1 Chemical Hazards 70
4.2.5.2 Glassware Hazards 71
4.2.5.3 Laser Light 72
4.2.5.4 Fire Hazards 72
4.2.6 Hazard Labeling: National Fire Protection Association 73
4.2.7 Summary of Important Safety Rules 75
4.2.8 Safety in Teaching Laboratories 76
4.2.9 Evacuation Procedures 78

4.3	SCIENTIFIC REPORT WRITING	78
	4.3.1 Introduction	78
	4.3.2 Getting Started	79
	4.3.3 Report Format	81
	4.3.3.1 Abstract	81
	4.3.3.2 Introduction	81
	4.3.3.3 Materials and Methods	82
	4.3.3.4 Results	82
	4.3.3.5 Discussion	83
	4.3.3.6 Conclusion	83
	4.3.3.7 References	83
	4.3.4 Closing Remarks	84
	REFERENCES	85

CHAPTER 5 ■ Nanotechnology Laboratories   87

5.1	SYNTHESIS OF GOLD NANOPARTICLES BY A WET CHEMICAL METHOD	87
	5.1.1 Aim	87
	5.1.2 Introduction	87
	5.1.3 Key Concepts	89
	5.1.4 Experimental	90
	5.1.4.1 Materials/Reagents	90
	5.1.4.2 Glassware/Equipment	90
	5.1.5 Special Safety Precautions	90
	5.1.6 Procedure: Preparation of Gold Nanoparticles by a Wet Chemical Method	91
	5.1.6.1 Part I: Concentrated Gold Nanoparticles with Sodium Citrate	91
	5.1.6.2 Part II: Concentrated Gold Nanoparticles without Sodium Citrate	92
	5.1.6.3 Part III: Dilute Gold Nanoparticles with Sodium Citrate	93

                5.1.6.4 Part IV: Dilute Gold Nanoparticles without
                          Sodium Citrate   94

        5.1.7   Characterization of Gold Nanoparticles   95

FURTHER READING MATERIAL   96

5.2   BIOSYNTHESIS OF ECO-FRIENDLY SILVER
       NANOPARTICLES   97

        5.2.1   Aim   97

        5.2.2   Introduction   97

        5.2.3   Key Concepts   99

        5.2.4   Experimental   99

                5.2.4.1 Materials/Reagents   99

                5.2.4.2 Glassware/Equipment   99

        5.2.5   Safety Precautions   100

        5.2.6   Procedure   100

                5.2.6.1 Part I: Biosynthesis of Silver Nanoparticles
                          Using Plant/Leaf Extracts   101

                5.2.6.2 Part II: Biosynthesis of Silver Nanoparticles
                          Using Green Tea   103

        5.2.7   Characterization of Silver Nanoparticles   104

FURTHER READING MATERIAL   105

5.3   SYNTHESIS OF ZINC SULFIDE NANOPARTICLES BY
       A REVERSE MICELLE METHOD   106

        5.3.1   Aim   106

        5.3.2   Introduction   106

        5.3.3   Key Concepts   109

        5.3.4   Experimental   109

                5.3.4.1 Materials/Reagents   109

                5.3.4.2 Glassware/Equipment   109

        5.3.5   Special Safety Precautions   109

        5.3.6   Procedure: Synthesis of Zinc Sulfide Nanocrystals
                via Reverse Micelle Method   110

                5.3.6.1 Part I: Preparation of the Primary Oil
                          Phase Slurry   111

5.3.6.2 *Part II: Preparation of the Solution Fractions* 112

5.3.6.3 *Part III: Preparation of Concentrated ZnS Nanocrystals* 113

5.3.6.4 *Part IV: Preparation of Diluted ZnS Nanocrystals* 114

5.3.7 Characterization of Zinc Sulfide Nanocrystals 115

FURTHER READING MATERIAL 115

5.4 SYNTHESIS OF FLUORESCENT CARBON NANOPARTICLES FROM CANDLE SOOT 116

5.4.1 Aim 116

5.4.2 Introduction 116

5.4.3 Key Concepts 117

5.4.4 Experimental 117

5.4.4.1 *Materials/Reagents* 117

5.4.4.2 *Glassware/Equipment* 118

5.4.5 Special Safety Precautions 118

5.4.6 Procedure 118

5.4.6.1 *Part I: Synthesis of Fluorescent Carbon Nanoparticles from Candle Soot* 119

5.4.6.2 *Part II: Separation of Fluorescent Carbon Nanoparticles Using the Thin-Layer Chromatographic Method* 121

5.4.7 Characterization of Carbon Nanoparticles 122

FURTHER READING MATERIAL 122

5.5 SYNTHESIS OF ZINC OXIDE NANORODS BY A MICROWAVE METHOD 124

5.5.1 Aim 124

5.5.2 Introduction 124

5.5.3 Key Concepts 125

5.5.4 Experimental 125

5.5.4.1 *Materials/Reagents* 125

5.5.4.2 *Glassware/Equipment* 125

    5.5.5   Special Safety Precautions   126

    5.5.6   Procedure: Preparation of ZnO Nanorods by a
          Microwave Method   126

          5.5.6.1   *Part I: Preparation of ZnO Solutions*   127

          5.5.6.2   *Part II: Synthesis of ZnO Nanorods by a
                  Microwave Method*   129

          5.5.6.3   *Part III: Observations of ZnO Nanorods by
                  Ultraviolet Light and Optical Microscopy*   131

    5.5.7   Characterization of Zinc Oxide Nanorods   132

FURTHER READING MATERIAL   132

5.6 SYNTHESIS OF BIMETALLIC NANOPARTICLES BY
     WET CHEMICAL METHODS   133

    5.6.1   Aim   133

    5.6.2   Introduction   133

    5.6.3   Key Concepts   134

    5.6.4   Experimental   134

          5.6.4.1   *Materials/Reagents*   134

          5.6.4.2   *Glassware/Equipment*   134

    5.6.5   Special Safety Precautions   135

    5.6.6   Procedure: Synthesis of Bimetallic
          (Fe@Au, Fe@Ag) Nanoparticles by a
          Wet Chemical Method   135

          5.6.6.1   *Part I: Synthesizing Fe@Au Bimetallic
                  Nanoparticles with Capping Agent*   136

          5.6.6.2   *Part II: Synthesizing Fe@Ag Bimetallic
                  Nanoparticles with Capping Agent*   138

    5.6.7   Characterization of Fe@Au and Fe@Ag Bimetallic
          Nanoparticles   139

FURTHER READING MATERIAL   139

5.7 SYNTHESIS OF POLYMERIC NANOPARTICLES BY
     A MODIFIED VERSION OF THE SPONTANEOUS
     EMULSIFICATION SOLVENT DIFFUSION METHOD   140

    5.7.1   Aim   140

    5.7.2   Introduction   140

5.7.3 Key Concepts   143

5.7.4 Experimental   143

    5.7.4.1 *Materials/Reagents*   143

    5.7.4.2 *Glassware/Equipment*   143

5.7.5 Special Safety Precautions   144

5.7.6 Procedure: Synthesis of PLGA Nanoparticles
by a Modified Version of the Spontaneous
Emulsification Solvent Diffusion Method   144

    5.7.6.1 *Part I: Preparing Polymer (PLGA) Solution*   144

    5.7.6.2 *Part II: Preparing the Surfactant
(Polyvinyl Alcohol)*   145

    5.7.6.3 *Part III: Preparing PLGA Nanoparticles*   145

5.7.7 Characterization of Polymeric Nanoparticles   146

FURTHER READING MATERIAL   146

5.8 NANOFORENSICS: FINGERPRINT ANALYSIS
STEPPING TOWARD THE NANOWORLD   147

5.8.1 Aim   147

5.8.2 Introduction   147

5.8.3 Key Concepts   149

5.8.4 Experimental   149

    5.8.4.1 *Materials/Reagents*   149

    5.8.4.2 *Glassware/Equipment*   149

5.8.5 Special Safety Precautions   150

5.8.6 Procedure   150

    5.8.6.1 *Part I: Preparation of Fingerprint Slides*   150

    5.8.6.2 *Part II: Preparation of Carbon Toner
Fingerprint*   151

    5.8.6.3 *Part III: Preparation for Superglue
Fingerprint-Fuming Process*   152

5.8.7 Characterization of Fingerprint Analysis   153

QUESTIONS   153

FURTHER READING MATERIAL   154

5.9 SYNTHESIS OF ALGINATE BEADS AND
INVESTIGATION OF CITRIC ACID RELEASE
FROM A NANOSHELL COATING OF POLYMER   155

    5.9.1   Aim   155

    5.9.2   Introduction   155

    5.9.3   Key Concepts   158

    5.9.4   Experimental   158

        5.9.4.1   Materials/Reagents   158

        5.9.4.2   Glassware/Equipment   158

    5.9.5   Special Safety Precautions   159

    5.9.6   Procedure: Synthesis and Drug Release Profile of
        Drug-Loaded Alginate Capsules   159

        5.9.6.1   Part I: Preparation of Solutions and
                Alginate Beads   160

        5.9.6.2   Part IIa: Formation and Encapsulation
                of Alginate Beaded Capsules   164

        5.9.6.3   Part IIb: Formation of Chitosan-Coated
                Alginate Beaded Capsules   165

        5.9.6.4   Part IIIa: Encapsulated Dye Release
                Studies of Alginate Beaded Capsules   165

        5.9.6.5   Part IIIb: Acid Release Studies from
                Encapsulated Alginate Beaded Capsules   167

FURTHER READING MATERIAL   172

5.10 SUPERHYDROPHOBICITY AND SELF-CLEANING
EFFECT OF A SURFACE   173

    5.10.1 Aim   173

    5.10.2 Introduction   173

    5.10.3 Key Concepts   175

    5.10.4 Experimental   175

        5.10.4.1 Materials/Reagents   175

        5.10.4.2 Glassware/Equipment   176

    5.10.5 Special Safety Precautions   176

5.10.6 Procedure: Superhydrophobicity and
Self-Cleaning Effect of a Surface 176

5.10.6.1 *Part Ia: Leaf Preparation on Microscope
Slides—Hydrophilic Leaves* 177

5.10.6.2 *Part Ib: Leaf Preparation on Microscope
Slides—Superhydrophobic Leaves* 178

5.10.6.3 *Part IIa: Contact Angle Estimation of
Hydrophilic Leaves Using Digital Pictures* 179

5.10.6.4 *Part IIb: Contact Angle Estimation of
Superhydrophobic Leaves Using Digital
Pictures* 180

5.10.6.5 *Part IIIa: Self-Replicating Properties of
Leaf Wax, Hydrophilic Leaves—Contact
Angle Estimation Using Digital Pictures* 180

5.10.6.6 *Part IIIb: Self-Replicating
Properties of Leaf Wax, Superhydrophobic
Leaves—Contact Angle Estimation Using
Digital Pictures* 181

5.10.6.7 *Part IVa: Self-Cleaning Experiment Using
Carbon Black Toner—Hydrophilic Whole
Leaf* 182

5.10.6.8 *Part IVb: Self-Cleaning Experiment Using
Carbon Black Toner—Superhydrophobic
Whole Leaf* 183

5.10.6.9 *Part Va: Self-Cleaning Experiment Using
Carbon Black Toner, Hydrophilic Leaves—
Microscope Slides* 184

5.10.6.10 *Part Vb: Self-Cleaning Experiment Using
Carbon Black Toner, Superhydrophobic
Leaves—Microscope Slides* 185

5.10.7 Contact Angle Estimation 186

FURTHER READING MATERIAL 186

5.11 SAMPLE ANALYSIS USING SCANNING ELECTRON
MICROSCOPY 187

5.11.1 Aim 187

5.11.2 Introduction 187

5.11.3 Key Concepts   189

5.11.4 Experimental   189

    5.11.4.1 Materials/Reagents   189

    5.11.4.2 Glassware/Equipment   189

5.11.5 Special Safety Precautions   190

5.11.6 Procedure: Sample Analysis Using Scanning
    Electron Microscopy   190

    5.11.6.1 Solid Samples   192

    5.11.6.2 Biological Samples   193

    5.11.6.3 Powdered Samples   195

    5.11.6.4 Liquid Samples   195

FURTHER READING MATERIAL   197

5.12 SAMPLE ANALYSIS USING ATOMIC FORCE
    MICROSCOPY   198

5.12.1 Aim   198

5.12.2 Introduction   198

5.12.3 Key Concepts   200

5.12.4 Experimental   200

    5.12.4.1 Materials and Equipment   200

    5.12.4.2 Glassware/Equipment   201

5.12.5 Special Safety Precautions   201

5.12.6 Procedure: Sample Analysis Using Atomic Force
    Microscopy   201

    5.12.6.1 Liquids   203

    5.12.6.2 Solids   203

    5.12.6.3 Powdered Samples   204

FURTHER READING MATERIAL   206

CHAPTER 6 ■ Nanotechnology and Nanoscience Projects   207

6.1 INTRODUCTION   207

6.2 NANOSYNTHESIS PROJECTS   209

6.2.1 Gold NPs (Laboratory 5.1)   210

6.2.2 Silver NPs (Laboratory 5.2)   211

6.2.3   QDs ZnS/Se NPs (Laboratory 5.3)   211

6.2.4   ZnO NPs (Laboratory 5.5)   211

6.2.5   Bimetallic Nanoparticles (Laboratory 5.6)   211

6.2.6   Nanopolymers (Laboratory 5.7)   212

6.2.7   Alginate/Chitosan Beads (Laboratory 5.9)   212

6.2.8   Superhydrophobic Surfaces (Laboratory 5.10)   212

6.3   NANOCHARACTERIZATION PROJECTS   212

6.3.1   Atomic Force Microscopy   212

6.3.2   Scanning Tunneling Microscopy   213

6.3.3   Field Emission Scanning Electron Microscopy   213

6.3.4   Transmission Electron Microscopy   213

6.3.5   X-ray Diffraction   214

6.3.6   Raman Spectroscopy   214

6.3.7   UV-Vis Spectroscopy   214

6.3.8   Thin-Layer Chromatography   215

6.4   NANOENERGY PROJECTS   215

REFERENCES   216

INDEX, 221

Foreword

THE IMPACT OF NANOTECHNOLOGY on the public imagination, scientific research, and industry, starting at the end of the twentieth century, has been nothing short of phenomenal and promises to be more far reaching than the Industrial Revolution of the eighteenth century. Fictional accounts of how nanotechnology may develop are found in many books and film scripts today, yet in the nano world, fact truly is stranger than fiction. Contrary to what our senses lead us to expect, fundamental properties such as melting point and color turn out to be strongly size dependent once particle sizes of 10 nm or less are achieved; the melting point of gold drops by up to 500°C, and the familiar gold color is replaced by a ruby red. Even more counterintuitive are the effects predicted by quantum theory and can be harnessed, for example, in a quantum computer.

A talk, "There's Plenty of Room at the Bottom," given by physicist Richard Feynman in 1959 inspired the beginnings of nanotechnology. Feynman considered the possibility of building structures through the direct manipulation of individual atoms. Feynman was awarded the Nobel Prize in Physics in 1965, and nanoscience has proved to be fertile ground for further Nobel Prizes in subsequent years. For example, the 1996 Nobel Prize in Chemistry was awarded to Robert Curl, Harold Kroto, and Richard Smalley for their discovery of fullerenes, nanometer-size geodesic structures made of carbon atoms; the 2010 Nobel Prize in Physics was awarded to Andre Geim and Konstantin Novoselov for their work on graphene; and the 2012 Nobel Prize in Physics was awarded to Serge Haroche and David Wineland for their work on individual quantum systems.

The course manual offers students the opportunity to experience directly some of the magic of the nano world through a series of well-designed laboratory experiments, yet given the sizes involved (a nanometer is 1 millionth of a millimeter), how can these experiments be truly "hands on"? The answer lies in two other scientific developments that were awarded Nobel Prizes.

In 1932, Werner Heisenberg was awarded the Nobel Prize in Physics for the creation of quantum physics, and, in 1986, the Nobel Prize in Physics was divided: half to Ernst Ruska for his design of the first electron microscope and half to Gerd Binnig and Heinrich Rohrer for their design of the scanning tunneling microscope. Quantum physics allows the nano world to be understood theoretically; the advanced microscopes allow nanostructures to be directly visualized and even touched.

A theme of nanotechnology that the course manual well illustrates is that Nature long ago realized the benefits of nanostructures, and there is growing interest in science and industry in the emerging field of biomimicry that exploits this. A well-known example is the "Lotus effect," by which nanoscale structures on the leaves of the lotus plant make them superhydrophobic, with the result that water droplets just run off, taking dirt particles with them; this effect is now used commercially to make self-cleaning windows. The author discovered a similar effect in a plant native to Australia and, moreover, showed that the nanostructure could be easily removed and deposited onto other surfaces. The combined toughness and lightness of bones result from nanoscale inorganic crystals embedded in an organic matrix; the author of this book has successfully built on this theme to develop artificial bone that is readily accepted into a living host. The author has also pioneered the uses of nanostructured substrates to direct the assembly of skin cells to form tissue. Thus, the material in the laboratory manual comes from someone who is not only knowledgeable in the field but also is one of its pioneers.

The laboratory manual is an excellent and timely addition to the resources available to teach undergraduate and postgraduate students working in the nanosciences. Written by an internationally recognized expert in the field, the material is well presented, making nanotechnology accessible to students from a wide range of backgrounds. The scope covers nano-related topics in physics, chemistry, biology, and materials science and includes underlying theory as well as many hands-on experiments that have all been well tested and refined over the years this course was taught by the author. The experiments are underpinned by comprehensive descriptions of nanomaterials and the very important techniques of sample characterization that reveal the full nature of the nano world. The presentation is visually attractive, with excellent illustrations.

Most important, the experiments are innovative and exciting, and the author's strong enthusiasm for his topic is evident throughout the text; I hope this will inspire the next generation of scientists and engineers to discover the joys of the nano world, both in Nature and made in the laboratory. This book is destined to become a much-loved, and well-used, classic.

Gordon M. Parkinson
Professor of Minerals and Energy Innovation
Foundation Director, Nanochemistry Research Institute
Curtin University
Perth, Australia

Preface

Nanoscience is truly an interdisciplinary field with its core participants of physics, chemistry, biology, and engineering. The wave of nanotechnology- and nanoscience-based discoveries in the last decade has accelerated the pace of new technological developments and applications. Worldwide, many countries have recognized the value and potential research dividends of nanotechnology- and nanoscience-based research in the form of new technologies that will underpin the next Industrial Revolution. For these new nano-based technologies to develop, a new generation of scientists and engineers needs to be trained in both the theoretical understanding and practical applications of this rapidly developing field. Although there are many theoretical-based nanotechnology and nanoscience textbooks available to undergraduates and graduate students, there are relatively few practical laboratory-based books that present a hands-on approach to many of the key synthesis techniques and processes currently used in nanotechnology and nanoscience. This was the motivation for writing this practical laboratory-based manual, which is specifically aimed at introducing undergraduate students to the exciting world of synthesizing their own nanometer-scale materials and structures and then analyzing their results using advanced characterization techniques.

This manual was born out of the necessity to deliver a variety of laboratory exercises designed to give undergraduate students practical hands-on skills to complement their theoretical studies in the Murdoch University nanoscience program. The manual presents a series of practical exercises that form a complete semester course. As nanoscience program chair, I developed and have presented each of the laboratory exercises included in this manual, which is aimed at the undergraduate-level student. However, it can also be used as an introduction for students at the graduate level. The only prerequisites the student needs to undertake the laboratory exercises are some basic laboratory skills and the capability to perform the laboratory procedures safely. Although the laboratory exercises were originally developed for a physics program, students from chemistry, biology, and engineering have also undertaken the course successfully and moved on to careers in industry and academia. Thus, the laboratory exercises appeal to a wide range of students and disciplines.

The manual has been arranged in six chapters. Chapter 1 introduces the student to the nanometer-scale world and presents some interesting examples of nanometer-scale materials and structures found in nature. Chapter 2 presents a range of nanometer-scale materials and synthesis processes used to produce them. Chapter 3 introduces some of the advanced characterization techniques used to examine nanometer-scale materials and structures. Chapter 4 has two parts; the first part discusses laboratory safety and the identification of potential hazards in the laboratory. Importantly, the second part focuses on preparing a scientific report and the need to record and present the results of your research. Chapter 5 presents the various laboratory exercises. Finally, Chapter 6 presents a series of projects that a student can undertake while collaborating with a mentor or supervisor.

I would like to acknowledge my Murdoch Applied Nanotechnology Research Group members, who have supported and encouraged the realization of this manual. I wish to offer my sincere thanks to Dr. Derek Fawcett for providing critical feedback on the chapters of the manuscript as they were being developed and his constant support in pushing back the frontiers of nanoscience. I am also grateful to Dr. Xuan Le and Dr. Brundavanam for their constant attention to detail and commitment during the preparation of the manual. In addition, I would like to give special thanks to Dr. Xuan Le for her valuable graphical skills in preparing the technical illustrations. I would like also to acknowledge the support and cooperation of the team at Taylor & Francis. In particular, I would like to thank Ms. Francesca McGowan, who helped to develop the layout style for the manual, and her assitant. I would also like to thank Mrs. Amy Blalock and Mrs. Amber Donley for the superb production of this edition.

I do not think that it is possible for a person with a family to write a book productively without a good deal of support. I would like to particularly thank my wife, Varny, for her constant support, encouragements, and patience along the way.

Dr. Gérrard Eddy Jai Poinern
Foundation Director
Murdoch Applied Nanotechnology Research Group
Senior Lecturer, Physics and Nanotechnology
School of Engineering and IT
Murdoch University
Perth, Australia

About the Author

Gérrard Eddy Jai Poinern holds a PhD in physics from Murdoch University, Western Australia. He graduated with a double major in chemistry and physics from Murdoch University. Currently, he is the director of the Murdoch Applied Nanotechnology Research Group based at Murdoch University and is a senior lecturer in physics and nanotechnology. He was the nanoscience program chair and developed the undergraduate nanoscience course. He also discovered and pioneered the use of an inorganic nanomembrane for potential skin tissue engineering applications, which led to a university spin-off company, Cellumina. His research interests include nanotechnology-based applications for environmental remediation, food security, biomimicry, photothermal applications and photovoltaics, and carbon technologies. His other interests are in the areas of drug delivery and medical treatments, such as in nerve repair, stroke treatment, and skin burns. He has received the Bill and Melinda Gates Foundation Global Grand Challenge in Health Innovation Award in 2010 for his work in the development of biosynthetic materials and their subsequent application in the manufacture of biomedical devices.

List of Abbreviations

AAO	anodic aluminum oxide
AFM	atomic force microscopy
CNT	carbon nanotube
CVD	chemical vapor deposition
DLS	dynamic light scattering
DNA	deoxyribonucleic acid
DOPA	3,4-dihydroxy-L-phenylalanine
DSSC	dye-sensitized solar cell
ITO	indium tin oxide
LOC	lab-on-a-chip
LPCVD	low-pressure chemical vapor deposition
MAP	mussel adhesive protein
MBE	molecular beam epitaxy
MEMS	microelectromechanical systems
MOCVD	metal organic chemical vapor deposition
MRI	magnetic resonance imaging
NEMS	nanoelectromechanical systems
PCS	photon correlation spectroscopy
PECVD	plasma-enhanced chemical vapor deposition
PEG	polyethylene glycol
PVC	polyvinyl chloride
PVD	physical vapor deposition
QELS	quasi-elastic light scattering
SEM	scanning electron microscopy

STM	scanning tunneling microscope
SWNT	single-walled nanotube
TEM	transmission electron microscopy
TLC	thin-layer chromatography
TOPSe	trioctylphosphine selenide
UV-Vis	ultraviolet–visible
XRD	x-ray diffraction

CHAPTER 1

The Nano World

1.1 INTRODUCTION

Research and development into nanotechnology is a rapidly growing field, with many governments and industries worldwide spending billions of dollars trying to unravel the reasons why matter at the nanometer scale behaves differently from matter at the bulk scale and how to capitalize on these novel properties for the betterment of humanity. For example, the Unites States, realizing the importance of this new scientific frontier, has even set up a federally funded institute to coordinate research activities in nanotechnology and nanoscience. There are many global challenges facing the world today. They range from finding sustainable renewable energy sources, reducing pollution, supplying uncontaminated drinking water, reducing the use of pesticides and herbicides, improving agricultural outputs to feed an increasing population, and predicted changes in world climate resulting from global warming. Nanotechnology and nanoscience have the potential to deliver solutions to many, if not all, of these global challenges.

Worldwide, many universities have taken on this challenge and have set up nanotechnology and nanoscience courses at the undergraduate and postgraduate levels. With this in mind, the manual was developed to present a set of comprehensive nanotechnology-based laboratory exercises designed to introduce, engage, and inspire science and engineering students from various disciplines. The manual assists students to perform simple, straightforward, and interesting experiments in a laboratory setting. It should be pointed out from the start that this manual's main focus is on the student acquiring experimental techniques and developing hands-on skills in a laboratory setting. The laboratory exercises presented in Chapter 5 are about synthesizing nanoparticles and nanostructures via a range of wet chemistry pathways and then using various advanced characterization techniques (discussed in Chapter 3) to analyze the nanometer-scale products produced by the students.

Furthermore, the manual does not follow the traditional chemistry laboratory format by which gravimetric measurements are made before and during the laboratory exercise. The majority of the chemicals and reagents are prepared prior to the laboratory class, ready for use, so that more time can be devoted to actually doing the experiments and understanding the processes and subsequent analysis. From my experience, students respond positively to the hands-on approach provided by laboratory experimentation and the use of advanced characterization techniques to analyze their own synthesized samples. The laboratory work also complements their understanding of the theory behind many of the concepts covered in theoretical lectures. The front end of this manual gives some of the theoretical aspects of nanomaterials and how these are synthesized (Chapter 2), as well as the basic principles behind nanocharacterization techniques (Chapter 3) used today. As for detailed discussions of the theoretical aspects of nanotechnology and nanoscience, Table 1.1 can guide students to some helpful nanotechnology and nanoscience textbooks and some relevant reviews. Students who have completed these laboratory exercises successfully are usually spurred on to explore other nanotechnology-related projects in other fields [1]. To support this enthusiastic student, Chapter 6 contains a series of more demanding and exciting projects that build from their own laboratory work performed in Chapter 5.

1.2 NANOTECHNOLOGY

Today, the *nano* label is associated with many terms in both the scientific and nonscientific worlds. The first person to identify and discuss the nanometer-scale world was physicist Richard Feynman. In his famous 1959 Caltech talk, "There Is Plenty of Room at the Bottom" [Ref 2], he predicted that one day it would be possible to assemble structures atom by atom or write the whole of *Encyclopaedia Britannica* on the head of a pin. However, it was Norio Taniguchi from the University of Tokyo who stated the original definition of nanotechnology in 1974. He described it as "the processing of separation, consolidation, and deformation of material by one atom or one molecule" [Ref 3]. Nanotechnology currently can

TABLE 1.1 Background Literature Information about Nanoscience, Nanotechnology, and Other Related Fields

Area	Reference
Introductory nanotechnology	Poole, C. P.; Owens, F. J. *Introduction to Nanotechnology*; Wiley: New York, 2003.
Nanotechnology revolution	Wolf, E. L.; Medikonda, M. *Understanding the Nanotechnology Revolution*; Wiley: New York, 2012.
Nanotechnology industry trends	Schulte, J. *Nanotechnology: Global Strategies, Industry Trends and Applications*; Wiley: New York, 2005.
Introductory nanoscience and nanotechnology	Binns, C. *Introduction to Nanoscience and Nanotechnology*; Wiley: New York, 2010.
Nanoparticles	Masala, O.; Seshadri, R. Synthesis routes for large volumes of nanoparticles. *Annu. Rev. Mater. Res.* **2004**, *34*, 41–81.
	Albanese, A.; Tang, P. S.; Chan, W. C., The effect of nanoparticle size, shape, and surface chemistry on biological systems. *Annu. Rev. Biomed. Eng.* **2012**, *14*, 1–16.
	Zhang, R.; Khalizov, A.; Wang, L.; Hu, M.; Xu, W. Nucleation and growth of nanoparticles in the atmosphere. *Chem. Rev.* **2011**, *112* (3), 1957–2011.
Nanoparticles for cancer therapeutics	Steichen, S. D.; Caldorera-Moore, M.; Peppas, N. A. A review of current nanoparticle and targeting moieties for the delivery of cancer therapeutics. *Eur. J. Pharm. Sci.* **2013**, *48* (3), 416–427.
Metal and inorganic nanoparticles	Louis, C.; Pluchery, O. *Gold Nanoparticles for Physics, Chemistry and Biology*; Imperial College Press: London, 2012.
	Altavilla, C.; Ciliberto, E. *Inorganic Nanoparticles: Synthesis, Applications, and Perspectives*; Taylor & Francis: Boca Raton, FL, 2010.
	Johnston, R. L.; Wilcoxon, J. P. *Metal Nanoparticles and Nanoalloys*; Elsevier: New York, 2012.
Core/shell nanoparticles	Ghosh Chaudhuri, R.; Paria, S. Core/shell nanoparticles: classes, properties, synthesis mechanisms, characterization, and applications. *Chem. Rev.* **2011**, *112* (4), 2373–2433.
Introduction to nanomaterials	Vollath, D. *Nanomaterials: An Introduction to Synthesis, Properties and Applications*; Wiley: New York, 2013.
Nanomaterials, synthesis, and applications	Cao, G. *Nanostructures and Nanomaterials: Synthesis, Properties and Applications*; Imperial College Press: London, 2004.

Continued

TABLE 1.1 (*Continued*)　Background Literature Information about Nanoscience, Nanotechnology, and Other Related Fields

Area	Reference
	Hosokawa, M.; Nogi, K.; Naito, M.; Yokoyama, T. *Nanoparticle Technology Handbook*; Elsevier Science: New York, 2007.
	Schmid, G. *Nanoparticles: From Theory to Application*; Wiley: New York, 2011.
	Logothetidis, S. *Nanostructured Materials and Their Applications*; Springer: New York, 2012.
	Geckeler, K. E.; Nishide, H., *Advanced Nanomaterials*; Wiley: New York, 2009.
Nanocarbons	Sharon, M.; Sharon, M. *Carbon Nano Forms and Applications*; McGraw-Hill Education: New York, 2009.
	O'Connell, M. J. *Carbon Nanotubes: Properties and Applications*; Taylor & Francis: Boca Raton, FL, 2012.
Carbon nanotube synthesis	Prasek, J.; Drbohlavova, J.; Chomoucka, J.; Hubalek, J.; Jasek, O.; Adam, V.; Kizek, R., Methods for carbon nanotubes synthesis—review. *J. Mater. Chem.* **2011**, *21* (40), 15872–15884.
Nanocharacterization techniques	Kirkland, A.; Hutchison, J.; Chemistry, R. S. o. *Nanocharacterisation*; RSC: Cambridge, UK, 2007.
Raman spectroscopy	Zhang, S. L. *Raman Spectroscopy and its Application in Nanostructures*; Wiley: New York, 2012.
Nanotools	Hornyak, G. L. *Introduction to Nanoscience*; CRC Press: Boca Raton, FL, 2008.
Nanobiotechnology and nanomedicine	de la Fuente, J. M.; Grazu, V. *Nanobiotechnology: Inorganic Nanoparticles versus Organic Nanoparticles*; Elsevier: New York, 2012.
	Niemeyer, C. M.; Mirkin, C. A. *Nanobiotechnology: Concepts, Applications and Perspectives*; Wiley: New York, 2006.
	Xie, Y. *The Nanobiotechnology Handbook*; Taylor & Francis: Boca Raton, FL, 2012.
	Logothetidis, S. *Nanomedicine and Nanobiotechnology*; Springer: New York, 2012.

be broadly defined as "the modification, usage, knowledge and development of nanomaterials, nanotools, nanomachines and nanosystems in order to solve a problem or perform a specific action." Although it had been known for some time that submicron-scale materials displayed some unusual and exotic properties, it was not until the advent of the scanning probe microscope that the nanometer-scale world was truly discovered. In this new world, matter at the nanometer scale, (1–100 nm, where 1 nm is equivalent to 1×10^{-9} m) was found to have significantly different properties compared to those of the equivalent material at the macroscopic scale [Ref 4].

1.3 NANOSCIENCE

The interdisciplinary field of nanoscience can be considered a new frontier of science that investigates materials with unique properties and behaviors only found at the nanometer scale. To put the nanometer scale in perspective, an atom is around 0.2 nm in diameter; a red blood cell is approximately 7,500 nm in diameter. Table 1.2 presents a selection of naturally occurring things and man-made structures with their respective scale sizes to give an overall perspective of where the nanometer scale is compared to the macroscopic world.

TABLE 1.2   Comparative Size Scales between Man-made, Natural Objects and Some Physical Properties

Power of Ten	Prefix, Symbol	Examples	
		Natural Things	Man-made Objects
10^6	mega, M	Radius of Earth = ~6.366 Mm	—
10^3	kilo, k	Height of Mount Everest = 8.848 km	Span of the Golden Gate Bridge (distance between towers) = 1.280 km
10^2	hecto, h	Speed of sound in 1 second = ~343 m	Height of the Great Pyramid of Giza = 138.8 m
10	deca, da	Height of Niagara Falls = 52 m	Height of the Leaning Tower of Pisa = 55 m
10^{-2}	centi, c	Length of a very large mosquito = ~1.5 cm	Minimum diameter of a golf ball = ~4.3 cm
10^{-3}	milli, m	Length of average red ant = ~5 mm	Diameter of a pinhead = ~1–2 mm
10^{-6}	micro, μ	Size of red blood cell = ~7–8 μm	Microelectromechanical (MEMS) device width = ~10–100 μm
10^{-9}	nano, n	DNA diameter = 2.5 nm	Carbon bucky ball diameter = ~1 nm
10^{-12}	pico, p	Carbon–carbon bond length = 154 pm	—

Thus, as with any new frontier, there is considerable scope to investigate the properties of nanometer-scale materials (or nanomaterials) and the mechanisms governing their reactions and behavior with matter. For instance, we can easily quantify the flow of electrons in bulk metallic gold wire using Ohm's law, but the exact laws governing the electron flow in nanometer-scale gold wires (or nanowires) need further investigation to fully explain the experimental results so far. Therefore, nanoscience can be considered the study of materials, devices, structures, and properties at the nanometer scale, and it has the potential and opportunities to deliver many novel discoveries in this rapidly developing field.

1.4 NATURE'S BIOLOGICAL PATHWAY

There has been and there are still considerable discussions about the efforts made by various scientific groups and industries worldwide in discovering, producing, and manufacturing synthetic nanomaterials for the betterment of humanity. However, it should be pointed out that *nature* has been creating both nanomaterials and a diversity of amazing nanometer structures for many millions of years. For example, nature has engineered our genetic inheritance by constructing the building blocks of life composed of deoxyribonucleic molecules, which are around 2.5 nm wide and several micrometers in length. Furthermore, the magnificent iridescent blue and green colors of a peacock's feather are the result of daylight interacting with nanometer-scale structures found in the bird's feather. In both examples, nature has been able to engineer a spectacular nanometer structure that has achieved impressive functionality. It is nature's ability to engineer at the nanometer scale that has attracted many researchers to investigate various biological processes found in plants and animals in the hope of mimicking the formation and self-assembly mechanisms of nanomaterials. Many scientists and engineers believe that biologically mimicry (biomimicry) of natural biological processes will lead to many new and novel discoveries for the synthesis of nanomaterials and nanostructures, which can be subsequently exploited for the benefit of humanity. The following sections discuss some really amazing examples of nature's ability to engineer at the nanometer scale in a variety of biological systems and potential applications in a numbers of biomedical areas, such as drug delivery and medical devices.

1.5 EXAMPLES OF NANOMATERIALS AND NANOSTRUCTURES FOUND IN NATURE

In spite of the considerable efforts made in chemistry, physics, and biology to engineer advanced functional nanomaterials, progress to date has not being able to achieve the same level of diversity and functionality found in nature. The primary building blocks of nature are atoms and molecules, and nature has been able to manipulate them many times to create intricate materials, structures, and mechanisms that continually contribute to life on Earth. Science has studied nature for centuries, but it has only been in the last few decades that science has had the extensive capability to study materials and structures at the nanometer scale via advanced characterization tools such as atomic force microscopy (AFM), scanning tunneling microscopy (STM), transmission electron microscopy (TEM), and scanning electron microscopy (SEM). These characterization techniques, which are discussed in Chapter 3, have enabled scientists to study and unravel many of nature's amazing achievements. The following three examples illustrate the remarkable materials and structures that nature has been able to engineer from the nanometer scale up to the macro scale to enhance the survivability of the organism in its respective environment. Importantly, from the nanotechnology point of view, nature's remarkable solutions have the potential to be used in a variety of other devices and applications.

1.5.1 The Beak of the Humbolt Squid (*Dosidicus gigas*)

To eat, the Humbolt squid (*Dosidicus gigas*), like its octopus cousin, uses its extremely strong beak to kill, fracture, and dismember the hard shells protecting giant crabs and other sea creatures that are its prey. This extremely strong beak, with a tip many times harder than human teeth, is an integral part of the squid's fleshy body. Interestingly, there is no dramatic change from the hard beak material to the softer fleshy mouth of the squid. This fascinating property was investigated by Miserez et al. and found to be the result of a gradual change in the amount of cross-linked structural proteins from the soft mouth to the hard and very sharp beak tip [2]. The nanometer-scale structural proteins in the beak form a unique composite that produces the transition from compliant to stiff, which allows the squid to tear into its prey without cutting and injuring itself in the process. This biologically engineered composite highlights some superior material properties that conventionally manufactured materials, such as metals, ceramics, and polymers, do not possess.

Inspired by the material properties of the squid beak, Fox et al. undertook a series of studies to biomimic the architecture and surface wetness properties to engineer nanocomposites with the potential for use in the manufacture of structural components and devices [3]. Building from their studies of sea cucumber skins, which are normally soft in water and hard when dried in air, the group treated cucumber skins with functionalized cellulose nanocrystals. When the skins were exposed to light, the level of cross-linking between the cellulose nanocrystals was found to increase proportionally with the level of light exposure. Thus, by adjusting the level of light exposure, the treated skin formed an engineered film that was hard at one end and soft at the other end. The gradual change in stiffness along the film's length was also enhanced by the film's wetness, with the overall result the production of an engineered film that mimicked the varying stiffness of the squid beak. From a biomedical point of view, the influence of wetness on stiffness has many potential applications, especially when you consider that the internal environment of the human body is wet, while the outside environment is dry. One direct potential application of wetness- and stiffness-dependent materials is in the treatment of diabetes. Millions of people worldwide, usually living a sedentary lifestyle, are affected by the disease, which results from a deficiency in insulin molecules that are needed to regulate the sugar levels in the body. Because of this deficiency, most diabetics need daily administered injections to maintain insulin levels in the body; however, there are a number of issues associated with the administration procedure. To alleviate these issues, an insulin pump was specially designed to deliver measured doses of insulin via strategically placed needles located on the body. In some cases, the needles create sores and infections, which can ultimately lead to device failure. It is believed that using biologically inspired properties, such as the varying gradient and material wetness found in the squid beak, it should be possible in principle to develop novel biologically compatible materials that will be able to prevent problems currently experienced in administering insulin via needles and make the procedure safer for the use of future generations.

1.5.2 The Beard of the Mussel (*Mytilus edulis*)

The mussel *Mytilus edulis* has evolved in the harsh environment of the seashore, where the impact of wave motion, tidal movements, and buoyancy forces and the need to be anchored place a variety of strong stresses on the mussel. In fact, the survival of this sea creature depends on a tethering structure composed of numerous threads called a byssus, which is generally known as the "mussel's beard," as seen in Figure 1.1a. In this harsh environment, most man-made materials would ultimately fail, but the mussel's beard continues to maintain a strong bond and a durable anchor point. It was the strong and robust nature of the mussel's beard that attracted scientific interests, first by Brown [4] and later by Waite and Tanzer in the 1980s. The later study revealed that the threads of the beard contained adhesive materials called mussel adhesive proteins (MAPs). These proteins were capable of producing a permanent adhesion between inorganic and organic substrates [5]. The mussel secretes these sticky proteins as soft adhesive glue from the soft foot of the mussel; this glue hardens within minutes to form a strong bond with a variety of substrates in water, where many man-made glues would fail to set and in turn fail to form an effective bond with the substrate. Biochemical studies of

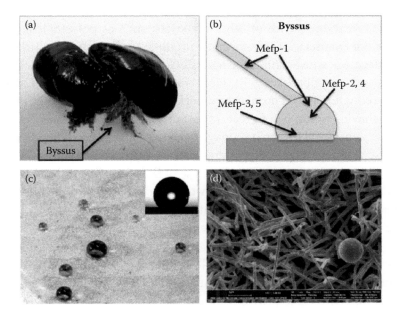

FIGURE 1.1   Nature's nanotechnology at play. (a) Mytilid mussel (top left) with fresh byssal thread for anchoring this sea mollusk to a hard surface. (Courtesy of J. Verduin, 2014.) (b) Schematic of the byssus foot and its adhesive pad with the relative distribution of Mefp proteins. (c) A superhydrophobic leaf from the rose of the west eucalyptus (*Eucalyptus macrocarpa*) from Western Australia. (d) High-resolution scanning electron image of the leaf's surface with nanometer-scale pillars around 280 nm in diameter.

the *Mytilus edulis* foot proteins (Mefps) present in the glue by Dalsin et al. found two key amino acid components: 3,4-dihydroxy-l-phenylanaline (DOPA) and hydroxyproline (Hyp) [6,7]. Studies to date have identified five Mefps, ranging from Mefp1 to Mefp5, and the mechanism used by the mussel to strengthen its attachment to the anchorage point, as seen in Figure 1.1b. During attachment, the mussel gradually changes the chemical composition of the beard's threads via the self-assembly of block copolymers (polymers from different monomers) made of collagen. Polymerization is catalyzed by metal ions such as zinc and copper ions, which are extracted from seawater during filter feeding by the mussel.

Further biological processes and materials that have potential applications in medicine can be found in the work of Vo-Dinh [8]. In particular, biological adhesives have many potential applications in both medical and dental procedures. For example, they could be used in procedures involving repairing mineralized hard tissues such as bone and teeth. Interestingly, studies of DOPA have revealed conflicting behavior toward proteins in the presence of other materials, such as polyethylene glycol (PEG). For example, DOPA normally promotes prolific protein sticking to its surface, but when conjugated with PEG, it repels protein biological molecules and even fibroblast cells. Fibroblast cell attachment studies showed a 95% reduction in cell numbers for periods up to 2 weeks. This cell reduction behavior would be of great importance in current medical implants, such as stents used to repair damaged arteries and artificial bloods vessels because these artificial components can be easily coated and eventually blocked by proteins and cells. Thus, use of biologically inspired materials with the potential to repel proteins would prolong the lifetime of medical implants in the body without the patient needing revision surgery to alleviate complications arising from occluded arteries and stents.

1.5.3 Superhydrophobic Properties Found in Some Plant Leaves

In the plant kingdom, there are many examples of nature engineering at the nanometer scale to produce some remarkable structures and materials to achieve unique properties, such as superhydrophobic surfaces. The most famous example is the lotus leaf (*Nelumbo nucifera*), which is an aquatic plant found in the equatorial regions of the world. The lotus has long been the symbol of purity in Asia and is used in many religious ceremonies. Its reputation for purity stems from its remarkably clean appearance in the muddy waters where resides. Interestingly, despite heavy rains during the day, its large flat leaves remain on the surface of the water and are not forced into the muddy water, where the plant's chances for photosynthesis would be significantly reduced. Through evolutionary pressures for survival, the lotus plant has been able to biologically engineer a leaf surface with microscopic bumps and nanometer-scale wax crystals. The combination of microscopic bumps and nanometer-scale wax creates a surface structure that makes the water bead instead of wetting the leaf surface. It is much like a fakir sleeping on a bed of nails.

Barthlott and Neinhuis, studying surface features of a variety of plants (among them was the lotus plant) using an SEM, observed micrometer-scale bumps covering the surface of the lotus leaf [9]. During these studies, they discovered further that the lotus leaf had the remarkable ability to self-clean. On a normal leaf, raindrops fall onto the leaf and then roll downward because of gravity. During the travel, the raindrop picks up contaminants and then deposits the contaminants back onto the leaf's surface at a different lower part of the leaf. However, in the case of the lotus leaf, the raindrop together with the contaminants continues to roll downward and in the process clears a pathway free of contaminants, self-cleaning the leaf's surface. This phenomenon has also been seen on taro (*Colocasia esculenta*) and kohlrabi (*Brassica oleracea*) leaves. When the water droplet has a contact angle greater than 150°, the leaf's surface is described as superhydrophobic or displaying the *lotus effect*. The concept of self-cleaning has many potential applications, ranging from nonwetting umbrellas to superhydrophobic channels on a lab-on-a-chip device for medical diagnosis and even sails capable of collecting water from the air.

The discovery of superhydrophobic surfaces and self-cleaning properties of several plant leaves has stimulated considerable research into producing specialized materials with superhydrophobic properties. Many groups worldwide are currently investigating natural materials for any potential superhydrophobic or superhydrophilic properties that could be biologically mimicked to manufacture materials with new and novel surface properties. Recently, I investigated the superhydrophobic and self-cleaning properties of an indigenous Australian desert eucalyptus plant (*Eucalyptus macrocarpa*), commonly called the Mottlecah or *rose of the west* (Figure 1.1c). During SEM examination of the leaf's surface, bumps of around 20 µm in diameter and regularly spaced at a distance of around 26 µm were identified. Closer examination revealed that the bumps were capped with nanometer-scale wax pillars with an average diameter of 280 nm at the tips (Figure 1.1d). Furthermore, water contact angle measurements revealed that the Mottlecah leaf was superhydrophobic, with an angle of 162° [10].

Interestingly, my team was able to remove the wax from the leaf and deposit it on a laboratory glass slide. The resulting wax coating self-assembled into the leaf's nanometer-scale pillars and was able to replicate the self-cleaning property of the Mottlecah leaf. After the deposition process, the normally hydrophilic glass surface was converted into a highly hydrophobic surface. Laboratory 5.10 provides the exciting opportunity to measure the water contact angle and investigate possible self-cleaning properties of some leaves because of their hierarchal micro-/nanostructures.

1.6 OVERVIEW OF CHAPTERS

Chapter 2 introduces nanometer-scale materials and the various synthetic methods and techniques used to manufacture these materials with nanometer-scale dimensions. Nanometer-scale materials such as metals, metal oxides, and quantum dots (QDs) are discussed along with the processing routes used to synthesize these materials. This is followed by Chapter 3, which looks at advanced characterization techniques such as AFM, STM, SEM and TEM. It is important to have a fundamental understanding of these techniques because many of the developments in nanotechnology heavily depend on the ability to identify, characterize, and analyze materials and structures at the nanometer scale. Because of the importance of characterization techniques, the chapter highlights the main techniques currently in use and briefly discusses their operating principles. Many of these techniques can be used to analyze the synthesized materials produced in each of the laboratory experiments and projects.

Chapter 4 has two parts; the first discusses laboratory safety, and the second discusses how to write proper scientific reports. Knowing how to safely handle chemicals and work safely and efficiently in a laboratory environment, whether in an academic setting or in industry, is important for today's scientists and engineers. The various key aspects of laboratory safety are discussed, and the importance of following laboratory rules is emphasized. The remaining part of the chapter focuses on writing meaningful experimental reports that effectively convey the results of the laboratory work in a clear, coherent, and logical style that can be easily understood by its readers.

Chapter 5 contains all the laboratory experiments, which are designed either to study a particular synthesis technique to produce a nanometer-scale material (e.g., gold nanoparticles) or to develop skills in using advanced characterization techniques to investigate nanometer-scale materials or structures. Following from Chapter 5, Chapter 6 presents a series of advanced projects designed to develop further laboratory and characterization skills and explore other related properties of nanoscale matter or create devices based on nanoparticles and nanostructures.

REFERENCES

1. Richmond, W. R. Contextual chemistry and physics teaching in an undergraduate nanotechnology degree. *J. Mater. Educ.* **2008**, *30* (1/2), 23–28.
2. Feynman R., There is plenty of room at the bottom, *Engineering and Science,* (**1960**), *23* (5), 22.
3. Taniguchi N., On the basic concept of nano-technology, *Proc. Intl. Conf. Prod. Eng. Tokyo, Part II*, Japan Society of Precision Engineering, **1974**.
4. International Organization of Standardization: Nanotechnologies-Terminology and Definitions for Nano-Objects, ISO/TS 27687:2008(E). ISO, Geneva, Switzerland (**2008**).
5. Miserez, A; Schneberk, T; Sun, C. J.; Zok, F. W.; Waite, J. H. The transition from stiff to compliant materials in squid beaks. *Science* **2008**, *319* (5871), 1816–1819.
6. Fox, J. D.; Capadona, J. R.; Marasco, P. D.; Rowan, S. J. Bioinspired water-enhanced mechanical gradient nanocomposite films that mimic the architecture and properties of the squid beak. *J. Am. Chem. Soc.* **2013**, *135* (13), 5167–5174.
7. Brown, C. Some structural proteins of *Mytilus edulis*. *Q. J. Microsc. Sci.* **1952**, *3* (24), 487–502.
8. Waite, J. H.; Tanzer, M. L. Polyphenolic substance of *Mytilus edulis*: novel adhesive containing L-dopa and hydroxyproline. *Science* **1981**, *212* (4498), 1038–1040.
9. Dalsin, J. L.; Hu, B.-H.; Lee, B. P.; Messersmith, P. B. Mussel adhesive protein mimetic polymers for the preparation of nonfouling surfaces. *J. Am. Chem. Soc.* **2003**, *125* (14), 4253–4258.
10. Dalsin, J. L.; Lin, L.; Tosatti, S.; Vörös, J.; Textor, M.; Messersmith, P. B. Protein resistance of titanium oxide surfaces modified by biologically inspired mPEG–DOPA. *Langmuir* **2004**, *21* (2), 640–646.
11. Vo-Dinh, T. *Nanotechnology in Biology and Medicine: Methods, Devices, and Applications*; Taylor & Francis: Boca Raton, FL, 2007.
12. Barthlott, W.; Neinhuis, C. Purity of the sacred lotus, or escape from contamination in biological surfaces. *Planta* **1997**, *202* (1), 1–8.
13. Poinern, G. E. J.; Le, X. T.; Fawcett, D. Superhydrophobic nature of nano-structures on an indigenous Australian eucalyptus plant and its potential application. *Nanotechnol. Sci. Appl.* **2011**, *4*, 113–121.

CHAPTER 2

Nanomaterials and Their Synthesis

2.1 INTRODUCTION

The study of materials at the nanometer scale (typically 1–100 nm) was spurred on by the realization that the properties of these materials have a strong dependence on their inherent shape and size [1]. Size, shape, surface structure, and chemical composition are all parameters that can be varied during synthesis to create nanometer-scale materials with novel properties that have the potential to be used in a variety of novel applications.

Materials can be made from several types of elements, either in the pure elemental form or in the form of compounds and composites. Generally, bulk materials can be classified broadly as metals, semiconductors, and insulators. And when any of these materials is produced in the nanometer scale, each displays shape-/size-dependent properties. These new properties have the potential to provide enormous opportunities for both scientists and engineers to create many novel applications that are normally not possible with conventional bulk materials. Many of the nanometer-scale properties (e.g., size, shape, surface structure, and chemical composition) have only been deciphered since the advent of advanced microscopic techniques, which has enabled researchers to precisely measure and directly visualize materials at the atomic scale in real time, something impossible just a few decades ago. And, even more impressive is the ability of these characterization techniques to give us a glimpse of materials and processes in their own localized micro-/nanoscopic environment. For example, techniques based on scanning probe microscopy can be applied to study the electrolytic corrosion of aluminum by chloride ions at the atomic scale in real time.

One amazing feature of using techniques such as scanning tunneling microscopy (STM) and atomic force microscopy (AFM) is the ability to use the probes to manipulate matter at the atomic scale. For example, STM was used by Eigler's group at IBM to position 35 xenon atoms on a single crystal of nickel at low temperatures and under ultrahigh vacuum to spell out the IBM logo over a period of 22 hours [2]. Since then, several groups have successfully used atoms such as iron (Fe) and molecules such as carbon monoxide to create surface features not previously seen in nature. These types of probe microscopy techniques have clearly demonstrated their capability of engineering at the atomic scale, and it makes them an ideal tool for producing nanometer-scale surface structures never seen before and to map out their inherent properties.

With these aspects in mind, this chapter provides relevant background information regarding a variety of nanometer-scale materials. Five groups of nanometer-scale materials are discussed: metals, metal oxides, polymers, quantum dots (QD), and nanocarbons. The remainder of the chapter looks at some commonly used synthesis techniques for producing nanometer-scale materials.

2.2 NANOMETER-SCALE MATERIALS

The properties of materials at the nanometer scale are strongly correlated to the size and shape of the particular element or compound. Materials in their bulk form have significantly different properties from those they have at the atomic scale. At this scale, atoms, molecules, and assemblies of these atoms and molecules are dominated by quantum effects. However, as the atoms and molecules assemble to produce matter at the macroscopic scale, the very large numbers of integrated atoms and molecules give the material its bulk properties. It is during the assembly process that the relationship between particle size and surface area becomes apparent. This important relationship can significantly contribute to properties such as optical response, surface plasmon resonance, and nanocatalytic properties of the material. The relation between size and surface area can be exemplified by the following exercise, which is in a sense similar to how the ancient Greek Democritus devised the atomic structure of matter: Consider a cube with 1 m sides; the volume would be 1 m^3, and the surface area of the six sides comes to 6 m^2. If this cube side is halved horizontally and vertically, 8 cubes would be generated, each with a side dimension of 0.5 m. In this case, the 8 cubes will have a total surface area of 12 m², and the next iteration involving 64 cubes will produce a total surface area of 24 m². As this procedure continues, the total surface area increases exponentially; by the fifth iteration, there are 32,768 individual cubes, each with a 0.03125-m side as seen in Figure 2.1a.

Another way to highlight the extreme surface area of smaller particles can be seen in the following example: Consider how many cubes with a side of 1 nanometer (nm) could be produced from a parent cube with 1-m sides and the resulting surface area created in such a process. In this case, the volume $(1 \times 10^{-9})^3$ of the nanocube would be 1×10^{-27} m³. The surface area of each single nanocube would be $6 \times (1 \times 10^{-9})^2 = 6 \times 10^{-18}$ m², with a phenomenal $(1$ m³$/10^{-27}$ m³$) = 10^{27}$ nanocubes produced. This means that the total surface all the cubes is $10^{27} \times [6 \times 10^{-18}$ m²$] = 6 \times 10^9$ m². This is equivalent to 6000 km², which is the land area of some states or even countries (in comparison, the land area of Puerto Rico is about 8870 km²). Taken in this context, it gives you an idea how powerful nanotechnology can be, especially when you consider that the catalytic power of a particular element is directly related to its surface area. Another interesting feature of nanometer-scale materials is that the proportion of surface atoms is large compared to bulk forms of the material, for which the surface atoms

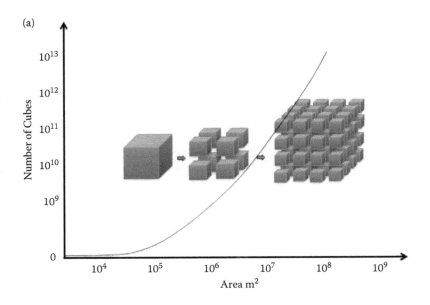

FIGURE 2.1   (a) Relationship between the number of cubes from a 1-m cube and the surface area produced.

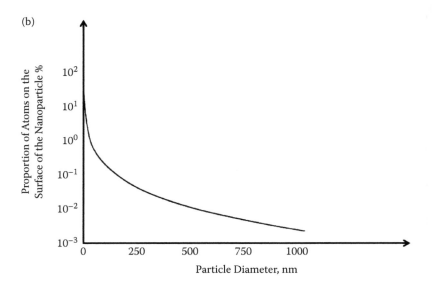

FIGURE 2.1 (*Continued*) (b) Relationship between the proportion of NP surface atoms versus the particle diameter.

only form a thin film above the core atoms. This is graphically presented in Figure 2.1b, in which the proportion of surface atoms is plotted against particle diameter. There is also an inverse trend seen in the surface area, which rapidly decreases with increasing particle size. Both the number of surface atoms and the surface area dependence on particle size also influence the surface adsorption capacity of material, which is an important factor in chromatography and separation technologies. For example, some activated carbons can have surface areas greater than 1000 m^2 and yet weigh just a few grams. Although every element and compound has the potential to be engineered at the nanometer scale, few elements and compounds have been investigated. Section 2.3 discusses several major types of nanometer-scale materials that have been investigated because of their characteristics, interesting properties, and possible applications.

2.3 TYPES OF NANOMETER-SCALE MATERIALS

The first indications that nanometer-scale materials had different and remarkable properties originated with the discovery of buckminsterfullerene (C_{60} or buckyball) in 1985. Subsequent studies ultimately led to the discovery of several other forms of exotic carbon structures, such as carbon nanotubes (CNTs; both single wall and multiwall), intercalated

CNTs, carbon nanohorns, and recently graphene. These discoveries spurred researchers worldwide to actively investigate other nanometer-scale materials, especially those with inherently novel properties that could be proprietarily secured via trademarks and patents.

In this pursuit, every synthesis process has been revisited, optimized, and refined to produce nanometer-scale materials. In addition, almost all of the elements listed in the periodic table have been scrutinized for potential production of nanometer-scale materials to reveal their potentially novel properties.

Today, the range of elements and compounds successfully synthesized in nanometer-scale forms, characterized, and even deployed as commercial products include the following:

- Metals

- Metal oxides

- Polymers

- Semiconductors

- Carbon compounds

However, there are still considerable research efforts worldwide into unlocking the intricate mechanisms involved in synthesizing nanometer-scale materials and mapping their unique properties. Effectively controlling the synthesis process to produce materials in the nanometer scale with great consistency is not an easy task. For example, some synthesis techniques may not be suitable for particular elements and may need to be modified or an alternative technique used. Another interesting feature of nanometer-scale materials occurs when they are incorporated into bulk materials. The resulting composite materials usually have enhanced properties, such as strength, durability, and tensile strength, compared to their bulk equivalents. For example, when small percentages of CNTs are added to polymers, there are significant improvements in strength in the overall doped polymer. In the following sections, major groups of nanometer-scale materials are briefly discussed; then, a range of commonly used synthetic techniques is outlined.

2.3.1 Nanometer-Scale Metals

Humanity has been using metals for many centuries, especially coinage metals such as gold (Au), silver (Ag), and copper (Cu). The earliest use of nanometer-scale metal particles was in the fourth century AD by the Romans, who incorporated them into their glass artifacts. A good example of this practice is found in the Lycargus Cup, which contains 70 nm Au and Ag nanometer-scale particles (NPs). So, it is not too surprising to find that there is currently considerable research into synthesizing nanometer-scale metals, investigating their properties and finding potential applications for these novel materials. When metals are synthesized at the nanometer scale, they exhibit properties not normally found in the micrometer scale or in the bulk form. The nanometer-scale forms exhibit quantum effects, lower melting temperatures, increased catalytic activity, and faster rates of reaction. The following sections discuss some examples of nanometer-scale metals (nanometals), their properties and their diverse range of applications.

2.3.1.1 Nanogold

The nanometer-scale properties of Au are significantly different from bulk properties of Au. For example, the melting point of bulk Au is 1064°C; 5-nm Au NPs display a reduced melting point of around 600°C. The observed difference in melting points has led to the manufacture of Au NPs coated with specific molecules designed to attach to cancerous cells. Then, the Au-loaded cancerous cells are targeted using a low-power laser light, which destroys these cells. This technique was successfully employed by El-Sayed's team at the Georgia Institute of Technology in 2006 to kill malignant cells loaded with Au nanorods targeted by a red laser (800 nm) [3]. One advantage of this technique is that only the cancer cells receive the largest Au loading; the surrounding cells with much lower loading are less affected by the treatment. Another application that takes advantage of the reduced melting temperature of nanogold was developed by Caruso's group at Melbourne University in Australia [4]. In this application, micron-size polymer capsules are filled with an anticancer drug and the final capsule coating containing 6-nm Au NPs. Thus, by irradiating these Au-sensitized capsules with a 10-ns pulse from a near-infrared laser, the nanogold melts and the capsules burst, releasing their drug payload. This is a typical example of how fundamental nanotechnology-based research into the properties of Au NPs has been able to be developed into a clever medical procedure for delivering anticancer drugs; it is all based on the reduction of the melting point temperature of this biocompatible metal.

For centuries, Au metal was considered completely inert; however, in 1989 Haruta et al. were able to demonstrate that nanometer-scale forms of

Au had catalytic activity and were able to catalyze chemical reactions [5]. In addition, the optical response of Au clusters varying in size, each containing thousands of atoms, produces beautiful colors, ranging from bluish tones to deep ruby red colors. Each of the colors is dependent on the capping or stabilizing agent used and on the size and shape of the Au NP. The colors result from interaction between the oscillating electric fields of the visible light and the free electrons located on the surface of the Au NPs. This creates a combined resonance oscillation known as the surface plasmon resonance. This causes the Au NPs to absorb light in the blue-green part of the electromagnetic spectrum, which causes red light to be reflected. This explains the rich red color of Au solutions. However, as Au NPs increase in size, the surface plasmon is red shifted toward longer wavelengths. Hence, red light is absorbed and as a consequence blue light is reflected. Thus, the larger Au NPs are more bluish or purple in color. For example, Palomba and coworkers from Birmingham University in the United Kingdom have compared the plasmon resonance of 7-nm Au NPs with that of 3-nm Au NPs and found that a blue shift was detected as the Au NP size was reduced [6]. These types of investigations are important because the size and shape of the Au NPs and their local environment can directly influence optical properties such as the local plasmon resonance. The influence of the local plasmon resonance on techniques such as dynamic light scattering and Raman spectroscopy is discussed in Chapter 3.

2.3.1.2 Nanosilver
Silver is the most electrically and thermally conductive metal known, and its antimicrobial activity has been known for centuries. However, it has only been shown recently that the nanometer-scale form of Ag can be incorporated into commercially available products such as Band-Aids, wound dressings, and clothing items like socks and garments to prevent bacterial/fungal growths. The Food and Drug Administration (FDA, United States) has approved nano Ag as a biocide and has resulted in this type of Ag being used in medical products such as catheters to provide an effective protective barrier against microbes. As an alternative to antibiotics, the potential use of nano Ag against pathogenic microbes, which are continually adapting and developing resistance to man-made antibiotics, is of great medical interest; this should lead to further development in this field.

Nano Ag in the form of Ag halides was first used in photography because Ag salts are extremely sensitive to light and can be used to generate photographic images. Prior to the digital age of compact electronic cameras with digital storage-and-retrieval systems, photographs were first photochemically formed on films, developed, and then printed on photographic paper. Photographic films were made from emulsions of Ag salt, which would release Ag atoms when a light ray entering the camera would strike the film. The agglomeration of Ag atoms to form nanoclusters would make the image visible. In the same way, photochromic glasses can be made to darken outdoors when high levels of ultraviolet (UV) light releases Ag atoms from the Ag salt embedded in the glass. The Ag atoms prefer to be with their own kind and in the process produce an Ag layer that is nanometers thick in the glass, which significantly reduces the UV and visible light levels entering the eye.

2.3.1.3 Nanocopper
Copper (Cu), like Au and Ag, is a conductive metal and is extensively used in cable wiring to carry electricity in electronic devices, computers, and electrical machines and is also used to bring electrical power to homes and industries. It is usually a red metallic solid, which can be made into nano Cu. The antibacterial and antifungal properties of nano Cu have the potential to be used in novel medical wound dressings and bandages. It has also been used in advanced coating if superior conductivity is required; another use us as an enhanced catalytic material in sulfide oxidation processes.

2.3.1.4 Nanoiron
Iron is one of the most-used engineering metals in the world today. It is the main constituent found in structural steels, which are used in the construction of buildings, ships, bridges, and railways. In the nano form, Fe behaves quite differently from its bulk form. For example, there is considerable research into using zero-valence Fe (Fe^0) or nano Fe to detoxify water contaminants and improve the quality of water for use in agriculture and drinking water supplies. There has also been extensive research into the interaction of nano Fe with electromagnetic radiation for potential medical applications. For example, engineered Fe NPs can be used to track the growth of cancerous cells via magnetic resonance imaging (MRI), which can provide early diagnostics of the disease. In addition, nano Fe can be used as a "Trojan horse" by rapidly building up in fast-growing cancerous cells, which can then be killed by induced hyperthermia-based therapies. This technique can deliver "pinpoint"

accuracy in targeting tumors because only highly nano-Fe-loaded cells and growths are destroyed.

2.3.2 Nano Metal Oxides

Nanometer-scale metal oxides have been intensely investigated over the last few decades. Like metals, metal oxides at the nanometer scale behave differently from their macroscopic forms. The members of the group of metal oxides consisting of Al_2O_3, TiO_2, ZnO, and Fe oxides all have interesting properties, which can be exploited in a number of nanotechnology-based applications. The following four sections provide a brief insight into these novel nanometer-scale materials and some interesting applications.

2.3.2.1 Aluminum Oxide

Aluminum oxide (Al_2O_3) is commonly called alumina and has a number of industrial applications, such as use as a catalyst support, an abrasive, an additive for advanced composites, and a paint additive. It is also used in the electronics industry because of its electrical insulation properties as a substrate for silicon (Si) for integrated circuit manufacturing. Nano forms of alumina can be used to produce scratch-resistant surfaces, and because of its transparent nature, it still allows light to pass through the surface film. It has also been used in some sunscreens and cosmetics.

Aluminum oxide layers produced on Al sheets via an electrochemical process called anodization can produce regular arrays of nanometer-scale pores. The growth of these pore structures can be controlled simply by adjusting such process parameters as voltage, electrolyte type and concentration, temperature, and time. By selecting the appropriate process parameters, the resulting anodic aluminum oxide (AAO) layer or membrane will have a pore structure with a high aspect ratio and a pore array that is highly regular. Importantly, the process is repeatable and consistently produces the desired nanometer-scale pore structure from the selected parameters. Because of this repeatability and the extreme regularity of the high-aspect pore structure, the membranes have been used as a template to manufacture regular high-aspect-ratio nanometer-scale structures such as nanorods, nanowires, and nanotubes inside the pore channels. From a biomedical perspective, these membranes are currently being investigated as a potential cell culture substrate for potential tissue engineering applications. Studies to date have shown that the regular surface structures of the AAO membrane have been able to solicit a positive response in terms of cell growth and proliferation for particular cell types [7].

2.3.2.2 Titanium Dioxide

Titanium oxide (TiO_2; also known as titania) powder is the main inorganic pigment added to paints in the world today. Micrometer-size powders are also widely used to whiten products such as paper and plastics. In addition, because TiO_2 is a biologically inert material, it is used in cosmetics and sunblocks and in the food industry, where is labeled as E171. Titania is a white micrometer-size powder; however, when the powders are made in the nanometer scale, they become transparent to visible light while still blocking UV light. This unique property has made nanometer-scale titania powders a prime candidate for a new generation of sunscreens and cosmetics. Nanometer-scale titania powders can be used as an effective self-cleaning coating on tiles and other types of surfaces that are prone to contamination. The titania coatings under sunlight tend to decompose organic buildups and effectively self-clean the contaminated surface.

Another interesting feature of nanometer-scale titania is its intrinsic semiconductor behavior, which makes it a suitable material for inclusion in novel dye-sensitized solar cells (DSSCs) or Grätzel cells. This relatively new type of solar cell is a viable alternative to currently expensive Si-based solar cells and has the potential to deliver a sustainable source of electrical power in the future.

2.3.2.3 Zinc Oxide

Micrometer-size zinc oxide (ZnO) powders have been used in a variety of sunscreens designed to prevent skin cancer because of its UV-blocking properties. The ZnO-based sunscreens are the first choice for many consumers because of their nontoxic properties and cost. The nanometer-scale form of ZnO consists of particles with dramatically larger surface areas, which offer both UV-A and UV-B protection against the sun's rays. In nanometer-scale formulations, ZnO is transparent and not only is suitable for sunscreens but also is even ideal for cosmetics worn every day.

2.3.2.4 Iron Oxides

The two main types of nanometer-scale iron oxide powders are ferric oxide (Fe_2O_3) and magnetite (Fe_2O_4). Ferric oxide NPs can be used in UV-blocking applications; magnetite NPs, as the name suggests, can be used in applications for which its magnetic properties can be utilized in enhanced storage of data for computer hard disks and other appliances.

Another application of magnetite or ferrofluids was developed by the National Aeronautics and Space Administration (NASA) in the 1960s for spacecraft development and technologies. In this application, a solution of NPs in suspension can interact strongly with electric/magnetic fields, and once the field is removed, the ferrofluid returns to its original liquid state. This allows for parts to be moved at will. The NPs are generally coated with a surfactant such as oleic acid and soy lecithin to keep them in suspension. Today, ferrofluids have been used in the medicine, electronics, materials science, and aerospace industries and their use continues to expand into other fields.

2.3.3 Nanopolymers

Polymers are an important class of materials and are the most extensively used materials in the world today. The word *polymer* is derived from the ancient Greek terms of *polis* (many) and *meros* (part), which suggests that these materials are made up of smaller units. Thus, polymers are large molecules made with up of many repeating subunits. Examples of industrially made polymers include polyethylene (PE; water bottles), polypropylene (PPE), polyvinyl chloride (PVC), and Teflon (polytetrafluoroethylene, PTFE). Examples of natural polymers include *Gutta percha* derived from the rubber tree (*Hevea brasiliensis*), cellulose from plant matter, polysaccharides, starch, and polypeptides. Polymer research is considered to have started with rubber manufacture in the nineteenth century following the Industrial Revolution. Subsequent developments by inventors such as Charles Goodyear in the rubber industry showed how the properties of natural rubber could be significantly improved by transforming the rubber into a semisynthetic polymer by the sulfur-based vulcanization process. However, it was not until 1907 that the first truly synthetic polymer called Bakelite was made by the Belgian American chemist Leo Baekeland. In this case, phenol and formaldehyde molecules were used as reactants to form a solid. Subsequent research revealed that polymers could be made from smaller constituent molecules, and the bonding process was through covalent bonding. The following is a short list of commercially available polymers currently in use:

- Nylon 66
- PE
- PTFE
- PPE

- PVC

- Polystyrene (PS)

- Polymethyl methacrylate (PMMA) acrylic

- Polylactide glycolic acid (PLGA)

- Polyhydroxyethl methacrylate (pHEMA)

Advancements in synthesizing biologically compatible polymers in the past few decades and the development of medical fields such as regenerative medicine and tissue engineering have seen a shift toward the use biocompatible polymers. One obvious advantage of using biocompatible polymers occurs during their degradation processes, with the gradual decomposition releasing nontoxic by-products. For example, sutures made from biocompatible PLGA degrade ultimately to lactic acid and glycolic components, which are easily handled and excreted from the body. Another area of considerable interest is pharmaceuticals, for which polymers can be designed to carry a payload of drugs. Polymer NPs can be designed and optimized to carry a wide variety of pharmaceuticals, which can be tailor-made to suit a particular type of disease, dosage, and release profile. Currently, a considerable amount of research is being undertaken by pharmaceutical companies in developing vaccines and anticancer drugs using drug delivery platforms based on nanometer-scale polymeric NPs [8]. Laboratory 5.7 gives you the opportunity to synthesize nanometer-scale polymeric particles of PLGA. Furthermore, Laboratory 5.8 allows investigation of the release characteristics from nano-/micrometer shells deposited onto alginate beads.

2.3.4 Quantum Dots

A quantum dot (QD) is a nanometer-scale crystal made from elements listed in groups II to VI and groups III to V in the periodic table. For example, cadmium selenium (CdSe) and zinc sulfide (ZnS) are two forms of QDs that are well known for their luminescent properties. QDs are classified as a (0D) semiconductor material, which means that it is confined in all three dimensions of space. Therefore, the motion of conduction band electrons, valance band holes, and excitons (pairs of conduction band electron and valence band holes) is confined within all three spatial directions. Although the QD term could be applied to metal clusters, it is generally reserved for only semiconductor materials. Furthermore, QDs can be considered fluorophores because they absorb light at a particular wavelength and then reemit the light at a different wavelength.

These nanometer-scale semiconductor crystals are used today in biomedicine as fluorescent markers because of their unique optical properties, which can be used to enhance images of cancerous growths and thus provide refined diagnostics for the surgeon and patient. Studies of QD properties have shown that the size of the QD energy band gap is directly dependent on the size of the QD. Thus, emitted wavelengths of light can be tuned from the UV range to the near-infrared (NIR) region by adjusting the size of the QD. For example, Bruchez et al. showed that 2-nm QDs will fluoresce in the blue region; 5-nm QDs emit in the red region [9]. Their studies confirmed that it is indeed possible to design materials with well-controlled optical characteristics.

Organic dyes have been extensively used as biological markers in the medical community, but there has been a gradual shift toward using CdSe QDs despite the toxicity Cd. However, if the CdSe QDs are encased in a protective ZnS shell to prevent Cd leaching, the QDs can be used safely. Because QDs have much greater photostability compared to the currently used organic fluorescent dyes, it is expected the use of QDs for imaging cells and cellular processes and in medical therapies will steadily increase in the future.

Another potential application of QDs is in the manufacture of novel solar cells, for which the ability to capture a larger proportion of the sun's spectrum using tunable NPs could yield much greater conversion efficiencies than current Si-based solar cells. For example, Murray et al. from Bawendi's group at the Massachusetts Institute of Technology have been able to manufacture QDs with a small energy band gap (0.4 eV) capable of absorbing light in the UV part of the spectrum all the way to the infrared (IR) region [10]. Importantly, they were able to process the QDs into a thin, lightweight film electrode, which has the potential to be used in flexible solar cells.

2.3.5 Nanocarbons

Carbon is a truly unique material; all life on Earth contains various forms of carbonic structures, from proteins to the tallest trees like the Californian redwood (*Sequoia sempervirens*). This is largely because of a process called catenation. In this process, an element can bond with itself to form long chains. It has been known for some time that there are two main allotropes of carbon: graphite and diamond. Graphite is black, soft, and conductive;

a diamond is shiny, transparent, and extremely hard. In terms of the bonding nature of the materials, the sp^2 hybridized bonding gives rise to two-dimensional layered structures like graphite, and the sp^3 hybridized bonding gives rise to diamonds. The diverse properties found in the different forms of carbon make this element remarkable and unique.

The first nanometer-scale forms of carbon to be discovered were the so-called buckminsterfullerenes or buckyballs, molecules that contain 60 atoms of carbon (C_{60}) and resemble soccer balls. The C_{60} structure is reminiscent of the geodesic dome designed by and structures created by the architect Buckminster Fuller in the 1940s. This discovery came about through an international collaboration of the United Kingdom and the United States, which used a high-powered laser to vaporize graphite samples and then analyzed the resulting vapor formed [11]. Analysis of the vapor revealed the presence of a novel carbon species with an atomic mass unit of 720 (60 × 12). This giant carbon molecule containing 60 atoms was found to be made up of 12 pentagons and 20 hexagons arranged in the exact shape of a soccer ball. The bonds between the carbon atoms are sp^2 in nature, and the diameter of the ball-like structure is 0.71 nm. Buckyballs can be crystallized into a face-centered cubic (fcc) lattice structure, with the crystal displaying semiconductor behavior with a band gap ranging from 1.5 to 2.0 eV.

Since the discovery of the buckyball, other fullerene molecules ranging from C_{70} all the way to C_{300} have also been found and added to the fullerene family. Some fullerenes have even been found in some minerals, meteorites, and even carbon soot [12]. Furthermore, the cage-like structure of the buckyball has also been used to encapsulate single atoms like La, Fe, Co, and Gd within its spherical structure to create endohedral metallofullerenes. Other types of fullerene reactions are also possible by which a hybrid molecule like $C_{59}N$ can be created. Today, this expanding field is making many exciting research discoveries, some of which are discussed in the following sections.

2.3.5.1 Carbon Nanotubes

Following the discovery of buckyballs in 1985, many research teams across the world started to search for new and novel nanometer-scale forms of carbon materials. For example, Ijima et al. discovered a tubular form of nanometer-scale carbons produced from an arc discharge machine [13]. The nanometer-scale needle-shaped structures were found growing from the negative electrode. Analysis of the needles revealed a coaxial graphitic sheet structure in the form of a tube, and the presence of helicity within the carbon tubes was also observed.

This new form of crystalline carbon in the form of tubes, with diameters as small as about 0.7 nm, is called a carbon nanotube (CNT). The tube structure can be either open or closed at the ends, and it can be several micrometers in length despite the diameter only being in the nanometer range. This geometry gives rise to a high-aspect-ratio material that can be greater than 100. CNTs have unusual chemical and physical properties that are very different from other carbon-based nanometer-scale structures. Some of the properties that CNTs display include low densities and exceptional tensile strength (>60 GPa), and they have extremely good electrical and thermal properties. For example, under relatively low electrical fields, closed CNTs have been found to be excellent electron emitters because of the presence of some sp^3 bonding at the tubes' apex. Generally, CNT structures are chemically stable, but they can also be functionalized with other molecules. Enhanced adsorption of gases and liquids has been observed in CNTs.

The simplest way to picture the structure of a CNT is to visualize one or more basal planes of graphite rolled up to form a series of closed coaxial cylinders. When one sheet of graphite is rolled up, the structure is called a single-wall carbon nanotube (SWNT); when more than one sheet is used, it becomes a multiwall carbon nanotube (MWNT). A CNT is described by its diameter d, its chiral angle θ, and the chiral vector $\mathbf{C} = n\mathbf{a}_1 + m\mathbf{a}_2$.

The chiral angle θ is defined as

$$\theta = \tan^{-1} \left[\frac{\sqrt{3}\,m}{m+2n} \right] \qquad (2.1)$$

The unit vectors $\mathbf{a}_1$ and $\mathbf{a}_2$ define a graphene sheet containing a hexagonal array of carbon atoms as presented in Figure 2.2. The (n, m) notation in the chiral vector is used to define the tube; the various carbon atom positions in the sheet can be seen, and ways to roll the sheet into a CNT along the tube axis can be seen. The image of the armchair orientation seen in Figure 2.2 gives rise to the concept of helicity as defined by the chiral vector.

A SWNT can be visualized as a graphene sheet wound around the outside surface of an invisible cylinder with an interatomic C–C distance (horizontal) of 14.4 nm (14.2 nm in graphite), with the armchair direction having a C–C distance of 28.3 nm (24.5 nm in graphite).

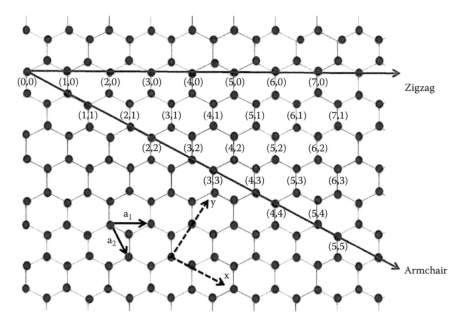

FIGURE 2.2   Hexagonal carbon lattice of the carbon nanotube.

Importantly, the CNT's diameter and chirality determine its overall electrical properties. For example, CNTs with the armchair-like structure have metallic properties, while both the zigzag and chiral structures give rise to semiconductor properties. SWNTs can be grouped into three classes: (1) zigzag nanotubes corresponding to (n, 0) or (0, m); (2) armchair nanotubes having (n, n); and finally (3) chiral nanotubes having generic values of n and m (n not equal to m). In the case of metallic SWNTs, the condition of n – m = 3k must be satisfied, where k is the greatest common divisor (or factor) of n and m. In the case of armchair nanotubes, these have metallic-like conduction properties; CNTs along other directions are semiconducting in nature.

2.3.5.2 Graphene

Graphene is a single layer of carbon atoms arranged in tightly bound hexagons, just one atom thick. This nanometer carbonic material can be generated easily from graphite where these are arranged in stacks. For instance, it would take about 3 million sheets of graphene on top of each other to make a 1 mm of graphite material. In 2004, teams including Andre Geim and Konstantin Novoselov from the University of Manchester, United Kingdom, demonstrated that graphene single layers could be isolated and tested, resulting in the award of the Nobel Prize for Physics in 2010. The increased worldwide interest in graphene molecules centers on its excellent mechanical, electrical, thermal, and optical properties. It is an extremely strong material with a fracture strength of about 125 GPa and a far superior thermal conductivity of about 5000 W $m^{-1}K^{-1}$; both of these properties would allow for the manufacture of novel superstrong material as well as the fabrication of enhanced material for conducting electrical charges in circuits and solar applications. Graphene has been shown to possess a large surface area (~2630 $m^2 g^{-1}$), and this can be capitalized on in the manufacture of supercapacitors for future energy storage devices. Thus, there is currently a rush into graphene and chemically modified graphene research and development to discover and exploit the wonderful and novel properties of these nanocarbons.

2.4 SYNTHESIS OF NANOMETER-SCALE MATERIALS

2.4.1 Introduction

In the fabrication and manufacture of NPs and nanometer-scale structures, several synthesis parameters must be effectively controlled to deliver a satisfactory outcome. Just being below the 100-nm mark is not enough, and the synthesis parameters must be controlled so that the following conditions are met: (1) Identical NPs are made every time (i.e., same diameter and shape); (2) they have the same morphology; (3) the same crystal and chemical bonding occurs whether on the surface or inside the NPs; and (4) the synthesis process must be stable. If these four conditions are met, then the synthetic process can be considered as reproducible and is a reliable technique. Today, there are many techniques capable of manufacturing NPs and nanometer-scale structures from solids, liquids, and gases. The solid base techniques used to manufacture NPs is straightforward and is usually done by attrition. Liquid-phase-based techniques include hydrothermal synthesis, coprecipitation, sol-gel processing, microemulsion, reverse micelle synthesis, microwave synthesis, ultrasound synthesis, and template methods. Gas-phase-based techniques are generally carried out by vaporizing a precursor material in a suitable atmosphere. This step is then followed by rapid cooling, which produces supersaturation and condensation to produce NPs and nanometer-scale structures.

In nanotechnology, there are two main methodologies used to design and manufacture NPs and nanometer-scale structures; the first is the top-down method, and the second is the bottom-up method. One popular top-down approach is via milling or attrition; a larger material is steadily milled down until NPs are produced. This technique is simple and straightforward and can produce NPs ranging in size from a few tens of nanometers to several micrometers. Unfortunately, the disadvantage of this technique is that it produces a wide range of particle sizes and morphologies. For this reason, particles produced by milling are limited to applications that do not require a specific particle size, shape, or size distribution. In addition, there can be significant levels of impurities produced during the process as a result of the milling medium. Another well-established top-down technique for manufacturing NPs and nanometer-scale structures is photolithography. This technique has been extensively used in the microelectronics and personal computer industries for fabricating Si-based integrated circuit chips and devices. Photolithography-based techniques have been able to meet the challenges of constant miniaturization of components down to the nanometer-scale level and still maintain fast throughput of components and devices today.

The bottom-up approaches are more widely used than top-down techniques such as photolithography, which requires a large capital investment in equipment and facilities. NPs and nanometer-scale structures can be made by either homogeneous or heterogeneous nucleation from liquids or vapors. For example, one widely used chemical method is to use micelles or reverse micelles to contain the chemical reactions with nanometer- or micrometer-size volumes. Within these confined volumes, nucleation and growth of NPs and nanometer-scale structures take place. The advantages of this technique are that it can be done at ambient conditions, and it can be easily scaled up to produce macroscopic quantities of nanometer material.

At this point, it is worth mentioning that the fabrication process can have a significant influence on the properties of the synthesized NPs and nanometer-scale structures. This means that each fabrication process will have inherent advantages as well as disadvantages, which will ultimately determine the quality and efficiency of producing the NPs and nanometer-scale structures. Therefore, selecting the appropriate fabrication technique to synthesize a specific nanometer-scale product with particular properties becomes an important factor that must be carefully considered. The following sections provide some background information and present several synthesis techniques specifically designed to produce particular nanometer-scale products.

2.4.2 Top-Down Techniques

2.4.2.1 Photolithography

Following the fabrication of the first transistor by Bardeen, Shockley, and Brattain in 1947, the development of transistor-based technologies has steadily increased. And, with the development of personal computers, mobile telephones, and similar electronic devices, the main driving force in the electronics industry is to continually strive for smaller components. Miniaturization of electronic components has resulted in the development of advanced photolithography-based techniques to produce integrated circuit chips that can literally contain many billions of transistors in a single chip. In this technique, UV light is shown onto a photomask after a surface treatment has been applied to the underlining substrate. This is followed by several chemical treatments to either etch a pattern into the surface of the substrate or deposit other materials. These processes are highly automated and expensive and carried out in clean rooms to reduce the risk of surface contamination. Integrated circuit chips with extremely small circuit patterns in the micrometer down to the nanometer range are made using this technique. The relatively new field of microelectromechanical systems (MEMSs) deals with the manufacture of very small components and devices between 1 and 100 μm in size. MEMSs have been used in a wide variety of applications, ranging from sensors (i.e., airbag sensors) to accelerometers used in touch telephones, laptop computers, and tablet devices. With advances and developments in nanotechnology, there is a gradual and inevitable move toward nanoelectromechanical systems (NEMSs). Currently, in the medical field there is extensive research into the development of the lab-on-a-chip (LOC) device. The overall size of these devices ranges from a few millimeters up to around a centimeter, and they can be mass produced at relatively low production costs, which makes them a single-use device. These devices are designed to collect body fluids and test them for a wide range of biological molecular signatures, thus providing early detection of particular diseases or illnesses. The embedded inner systems in the LOC device are made up of both micrometer- and nanometer-scale components; therefore, only small volumes of bodily fluids are sampled and tested. Because of the small volumes needed by the device, there is rapid analysis of the fluids, which in turn provides a rapid response for the medical personnel who are monitoring the patient.

2.4.2.2 Molecular Beam Epitaxy

The molecular beam epitaxy (MBE) technique is extensively used in the semiconductor industry to precisely deposit atomic layers of semiconductor materials onto a variety of substrates. Epitaxy is the process of growing a thin layer (epilayer) in an ultrahigh-vacuum system to prevent any contamination. In addition, a series of critically temperature-controlled ovens and precursor treatments is used during the deposition to ensure the correct stoichiometry of the layers. Both epitaxial layers and hetero-epitaxial layers can be deposited in this manner. Many quantum well devices, QDs, and similar nanometer-scale structures have been manufactured for telecommunications and personal computer industries using the MBE technique. The process is intensive and requires a substantial capital investment in expensive equipment and facilities.

2.4.3 Bottom-Up Techniques

2.4.3.1 Colloids

One of the most efficient and quickest ways to create metallic NPs is by using the aqueous reduction of soluble metallic complexes to create colloidal suspensions of NPs. The synthesized NPs have very high surface energies and will tend to agglomerate and form larger structures to reduce their surface energy. To prevent this and maintain a stable colloidal solution, a stabilizer in the form of a surfactant or an organic polymer is added to restrict particle growth and stop interparticle agglomeration. Therefore, the correct combination of low metal complex concentration and surfactant (usually called a capping agent) will efficiently produce monosize metallic NPs. Furthermore, to improve the output from this type of reaction process, various metallic complex precursors, reductants, and different growth media are available. Moreover, even using a two-step reaction instead of a single-step reaction can produce different types of NPs from the same initial raw materials.

Gold has fascinated humanity for many centuries, and since the middle Ages, it has been the alchemist's dream to convert base metals into Au. Even the mining and processing of Au has attracted considerable scientific and engineering interest over the centuries. In 1857, the famous experimentalist Michael Faraday published the results of a comprehensive study into the preparation of Au colloids. In fact, a sample of his nano Au is still on display in the British Museum. The most common method used to synthesize Au NPs was one pioneered by Turkevic et al. and involves the reduction of chloroauric acid ($HAuCl_4$) with sodium citrate at 100°C to produce an Au colloid [14]. In this case, the citrate molecules act as both reducing agent and capping agent to create a stable colloidal suspension. In Laboratory 5.1, you will have the opportunity to use a reduction process using sodium borohydride to create your own Au NPs. In a similar fashion, Ag NPs can be created when sodium borohydride is added to an aqueous solution of Ag nitrate. The borohydride molecule is a strong reducing agent, which readily reacts readily with the Ag^+ ions in solution to create Ag NPs. Here again, a capping agent is generally used to stabilize the colloid.

A recent and novel green chemical approach to synthesize NPs involves the use of natural biological molecules as reducing and capping agents. Plant extracts from leaves, stems, and roots have been used to synthesize a variety of metallic NPs, such as plates, rods, cubes, and even pyramids. In addition, both fungus and bacteria have shown the potential to synthesize NPs and appear to be low-cost and energy efficient ways to create NPs. In Laboratory 5.2, you will have the opportunity to use this green pathway to create your own eco-friendly Ag NPs.

2.4.3.2 Sol-Gel

Sol-gel materials are created when colloidal solution particles aggregate and form an interlaced network of linked particles, which are interpenetrated by the solution. The resulting degree of interconnecting within the gel determines whether it will be solid or semisolid. Sol-gels have successfully been used in the synthesis of NPs [15]. For example, a four-stage sol-gel method can be used to synthesize crystalline ZnO NPs, with zinc acetate dihydrate [$Zn(CH_3COO)_2.2H_2O$] used as the source of zinc ions. The four stages consist of solvation, hydrolysis, polymerization, and the final transformation to produce ZnO NPs. In the first and second stages, ethanol is used as a solvent to dissolve the zinc acetate dihydrate in the presence of monoethanolamine (MEA). The MEA molecule acts both as a base and as a complexing agent. In the third stage, the acetate ions are removed to produce a colloidal-gel-like material, and the ethanol molecules react to produce a polymer precursor with the zinc, which ultimately produces the ZnO NPs in the final stage. Alternatively, the final stage can be a heating process to form other ZnO nanometer-scale structures. The sol-gel method can easily be adapted to produce NPs of silica, iron oxides, and aluminum oxides. The advantages of this method include its straightforward procedure, inexpensive starting materials, and versatility.

2.4.3.3 Vapor Deposition and Chemical Vapor Deposition

Two popular vacuum-based techniques used to produce a wide variety of nanometer-scale materials and structures are physical vapor deposition (PVD) and chemical vapor deposition (CVD). During the PVD process, atoms or molecules from a vapor phase are deposited in thin layers onto a cooled substrate. In the CVD process, solid material from the vapor phase is deposited on reaction with a heated substrate. During the deposition, a chemical reaction takes place over the surface. Through time, the basic CVD process has been modified to produce NPs with particular properties using a variety of adaptations such as MOCVD (metal organic chemical vapor deposition), PECVD (plasma-enhanced chemical vapor deposition), and LPCVD (low-pressure chemical vapor deposition).

2.4.3.4 Sputtering

Sputtering is extensively used to prepare samples for electron microscopy. The technique involves depositing a conductive metal layer onto a nonconductive specimen to prevent electrical charge buildup. Typical metal targets such as Au and platinum (Pt) sheets are used as the source of the metallic coating. During the process, heavy and energetic argon (Ar) ions bombard the metal target and displace surface atoms; the metallic atoms are then directed toward the surface of the substrate. The thickness of the deposited film is controlled and can be as thin as a few nanometers. Furthermore, a wide range of compounds can be used as target material instead of the more conventional metallic target mentioned. The advantages of the technique come from its straightforward procedure and its ability to control the thickness of the layer during deposition.

2.4.3.5 Laser Ablation

In the laser ablation technique, a high-energy laser beam is used to vaporize a target so that the vaporized material can deposit onto the surface of a nearby substrate. The process is carried out within a vacuum chamber to avoid any potential reactions with atmospheric gases and prevents any unwanted reactions on the substrate. Because of the small size of the targeted area, the amount of ablated material that is deposited is also small. Therefore, it is possible to carefully control the deposition of compounds and in the process maintain an accurate stoichiometric balance if needed.

2.4.3.6 Anodic Aluminum Oxide Templates

Template-based synthesis techniques are a popular method for producing regular and identical nanorods, nanowires, and nanotubes. To produce these high-aspect-ratio nanometer-scale structures, a template with regular nanometer-scale channels is needed. Commercially available AAO membranes, and polymer membranes have been successfully used to produce high-aspect-ratio nanometer products.

However, many researchers have adopted the approach of making their own AAO membranes and in the process producing templates with predetermined geometries. The membranes are formed by the anodization of Al, in which an Al sheet acting as the anode of an electrical circuit is placed into an electrolytic bath containing a particular acid, such as oxalic, sulfuric, or phosphoric acid. The electrolytic bath is temperature controlled and, in the case of oxalic acid, maintained at 5°C. The anodization voltage is applied, and during the following hours, a disorganized and irregular oxide layer grows on the Al sheet. However, this porous oxide layer is unsuitable and is removed before a second anodization step is carried out. This two-step procedure was developed by Masuda et al. and is capable of producing regular hexagonal pores with densities as high as 10^{11} per square centimeter. The pore size distribution can range from a few nanometers to 500 μm, and it can be selected by controlling the acid type and concentration, temperature of the bath, and voltage applied across the electrodes [16].

The advantage of manufacturing in-house membranes is that only general laboratory equipment is needed to engineer a wide range of membranes. Following membrane fabrication, the nanometer-scale pore channels formed in the oxide can be used as nanosize test tubes to create a variety of materials. For example, the test tube can be filled with materials such as metals, semiconductors, and even polymers to manufacture regular nanowires, nanorods, and nanotubes, all with high aspect ratios.

Another nanotechnology-based application is to use the vapor liquid solid (VLS) process and the confined spaces of the pore channels to create nanowires, nanorods, and even CNTs [17]. In this case, a catalytic metallic element (Fe, Co, and Ni) is deposited at the base of the pore channel, and then the rods or tubes grow within the confined channels when a carbon gas is introduced and reacts. This template technique delivers rods, wires, and tubes all with the same length and diameter; in the case of carbon, it can deliver regular CNTs.

2.4.3.7 Spray Pyrolysis

The spray pyrolysis (or flame pyrolysis) technique can be used to fabricate nanometer-scale structures from materials such as oxides, which because of their refractory nature will only react at high temperatures. For example, if a metal catalyst needs to be deposited on a ceramic support, then a liquid precursor is made so that it can be sprayed. The spray can be produced using a low-power ultrasound device or atomizer, and then an inert carrier gas is used to carry the mist containing the precursors toward a flame burner. The reactants for the flame can be CH_4 and O_2 combusting in a set ratio to create a hot zone. In the flame, rapid processes such as solvent evaporation, drying, precipitation, and pyrolysis take place. As the processes continue, solid nanometer-scale material is produced and deposited on the ceramic support. This technique produces high-purity and high yields of NPs (grams per hour), which makes this method a cost-effective route for producing large quantities of NPs.

2.4.3.8 Ultrasonic Synthesis

Ultrasound-based technologies have emerged as a another powerful synthesizing technique used in nanotechnology-based research and applications. Most people would be aware of the use of ultrasound in imaging and medical diagnostics. As a form of radiation, ultrasound has the capacity to break chemical bonds and create products. The ultrasound frequency range is above 20 kHz and can reach frequencies in the gigahertz level. When a liquid like water is irradiated by ultrasound, bubbles are created, which rapidly grow and on reaching a maximum size, collapse (cavitation process). The sudden implosion of the bubbles (about 100 μm in size) creates a local hot spot, which can reach temperatures as high as 5000 K and pressures of up to hundreds of atmospheres depending on the level of ultrasound power used. At the same time, the cooling rates are extremely high and can be as great as 10^{11} K/s.

The capacity of ultrasound to create these extreme temperatures and pressures within a medium makes this technique attractive. For example, Suslick et al. were able to demonstrate that when $Fe(CO)_5$ was subjected to ultrasonic irradiation, NPs of Fe between 5 and 20 nm could be produced [18]. Furthermore, other researchers in this field have shown that several other types of nanometer-scale structures and NPs can be manufactured using this technique, such as QDs of CdSe and CdS and nanorods of magnetite, to mention just a few [19,20]. Because control of size and shape is of great importance in nanotechnology and nanoscience-based applications, size selection in ultrasound synthesis is accomplished by controlling the precursor concentrations. Studies have shown that there is a direct relationship between precursor concentration and overall size of NPs produced. Unfortunately, the shape and morphology are less predictable with this technique.

2.4.3.9 Microwave Synthesis

Microwaves are a form of electromagnetic radiation with frequencies between 300 MHz and 300 GHz and wavelengths between 1 m and 1 mm. Microwave ovens provide rapid heating compared to conventional conduction and convection heating systems. The rapid heating in this case is caused by the microwaves inducing the fast rotation of dipoles present in a material. This rapid rotation results in production of thermal energy, which heats the material. In fact, microwave frequencies correspond closely to the rotational excitation energies found in many materials [21]. For example, polar solvents heat easily throughout the whole volume. This is in direct contrast to the hot plate conventionally used in most chemical labs, where the thermal energy is transferred by conduction and thermal convections. Thus, in conventional heating there are large temperature gradients initially set up in the material medium during heating. Because microwave heating is extremely fast, the chemical rate of reactions is also extremely fast. Therefore, today there are many types of microwave ovens specifically designed to perform particular chemical reactions. Generally, a specially designed pressure-rated Teflon-lined vessel is used to contain the reactants to avoid any pressure buildup and possible explosions. In Laboratory 5.5 of Chapter 5 you will synthesize ZnO NPs on a glass slide using a microwave-oven-based technique.

REFERENCES

1. Theodore, L.; Kunz, R. G. *Nanotechnology: Environmental Implications and Solutions*; Wiley: New York, 2005.
2. Hornyak G. L.; Dutta, J.; Tibbals, H. F.; Rao, A. K. *Introduction to Nanoscience*; CRC Press: Boca Raton, FL, 2008, 34–35.
3. Huang, X.; El-Sayed, I. H.; Qian, W.; El-Sayed, M. A. Cancer cell imaging and photothermal therapy in the near-infrared region by gold nanorods. *J. Am. Chem. Soc.* **2006**, *128* (6), 2115–2180.
4. Radt, B.; Smith, T. A.; Caruso, F. Optically addressable nanostructured capsules. *Adv. Mater.* **2004**, *16* (23–24), 2184–2189.

5. Haruta, M.; Yamada, N.; Kobayashi, T.; Iijima, S. Gold catalysts prepared by coprecipitation for low-temperature oxidation of hydrogen and of carbon monoxide. *J. Catal.* **1989**, *115* (2), 301–309.

6. Palomba, S.; Novotny, L.; Palmer, R. Blue-shifted plasmon resonance of individual size-selected gold nanoparticles. *Opt. Commun.* **2008**, *281* (3), 480–483.

7. Poinern, G. E. J.; Shackleton, R.; Mamun, S. I.; Fawcett, D. Significance of novel bioinorganic anodic aluminum oxide nanoscaffolds for promoting cellular response. *Nanotechnol. Sci. Appl.* **2011**, *4*, 11–24.

8. Poinern, G. E. J.; Le, X. T.; Shan, S.; Ellis, T.; Fenwick, S.; Edwards, J.; Fawcett, D. Ultrasonic synthetic technique to manufacture a pHEMA nanopolymeric-based vaccine against the H6N2 avian influenza virus: a preliminary investigation. *Int. J. Nanomed.* **2011**, *6*, 2167–2174.

9. Bruchez, M.; Moronne, M.; Gin, P.; Weiss, S.; Alivisatos, A. P. Semiconductor nanocrystals as fluorescent biological labels. *Science* **1998**, *281* (5385), 2013–2016.

10. Murray, C. B.; Norris, D. J.; Bawendi, M. G. Synthesis and characterization of nearly monodisperse CdE (E = sulfur, selenium, tellurium) semiconductor nanocrystallites. *J. Am. Chem. Soc.* **1993**, *115* (19), 8706–8715.

11. Kroto, H. W.; Heath, J. R.; O'Brien, S. C.; Curl, R. F.; Smalley, R. E. C 60: buckminsterfullerene. *Nature* **1985**, *318* (6042), 162–163.

12. Bottini, M.; Mustelin, T. Carbon materials: nanosynthesis by candlelight. *Nat. Nanotechnol.* **2007**, *2* (10), 599–600.

13. Ijima, S. Helical microtubules of graphitic carbon. *Nature* **1991**, *354* (6348), 56–58.

14. Turkevich, J.; Stevenson, P. C.; Hillier, J. A study of the nucleation and growth processes in the synthesis of colloidal gold. Discuss. *Faraday Soc.* **1951**, 11, 55–75.

15. Znaidi, L.; Touam, T.; Vrel, D.; Souded, N.; Ben Yahia, S.; Brinza, O.; Fischer, A.; Boudrioua, A. ZnO thin films synthesized by sol-gel process for photonic applications. *Acta Phys. Pol. A* **2012**, *121* (1), 165–168.

16. Masuda, H.; Hasegwa, F.; Ono, S. Self-ordering of cell arrangement of anodic porous alumina formed in sulfuric acid solution. *J. Electrochem. Soc.* **1997**, *144* (5), L127–L130.

17. Cao, G.; Liu, D. Template-based synthesis of nanorod, nanowire, and nanotube arrays. *Adv. Colloid Interface Sci.* **2008**, *136* (1), 45–64.

18. Suslick, K. S.; Choe, S.-B.; Cichowlas, A. A.; Grinstaff, M. W. Sonochemical synthesis of amorphous iron. *Nature* **1991**, *353* (6343), 414–416.

19. Li, H.-l.; Zhu, Y.-c.; Chen, S.-g.; Palchik, O.; Xiong, J.-p.; Koltypin, Y.; Gofer, Y.; Gedanken, A. A novel ultrasound-assisted approach to the synthesis of CdSe and CdS nanoparticles. *J. Solid State Chem.* **2003**, *172* (1), 102–110.

20. Kumar, R. V.; Koltypin, Y.; Xu, X.; Yeshurun, Y.; Gedanken, A.; Felner, I. Fabrication of magnetite nanorods by ultrasound irradiation. *J. Appl. Phys.* **2001**, *89* (11), 6324–6328.

21. Rao, K.; Vaidhyanathan, B.; Ganguli, M.; Ramakrishnan, P. Synthesis of inorganic solids using microwaves. *Chem. Mater.* **1999**, *11* (4), 882–895.

Characterization Methods for Studying Nanomaterials

3.1 INTRODUCTION

The origins of science, its advancements, and its development have paralleled closely the advances and use of microscopy. The discovery of light microscopy in the second half of the 1600s gave science its first glimpses of cellular structures found in bacteria, plants, and animal tissues. In the following centuries, microscopy-based techniques significantly improved and, with continuing development, so did the understanding of the natural world. Today, science continues its exploration of the natural world, but science has new types of microscopic methods and characterization techniques that not only provide images but also can provide fundamental information regarding the composition and properties of materials all the way down to the atoms. In particular, nanotechnology owes much of its history and advancement to the development of new characterization tools such as the scanning tunneling microscope (STM) and the atomic force microscope (AFM), which are capable of not only imaging atoms but also manipulating them at will. And, there are other techniques, such as transmission electron microscopy (TEM) and scanning electron microscopy (SEM), that are used in imaging materials to determine their size, shape, composition, and morphology. The range and limits of these microscopy techniques are listed in Table 3.1. The microscopy-based techniques can be considered

TABLE 3.1 Chart of Microscopy and Type of Information Generated

Microscopy	Resolution Limit	Characteristics
Light microscopy	~0.2 μm	Samples can be imaged in liquid or air. Resolution is limited by the wavelength of visible light.
Fluorescent microscopy	~0.2 μm	Samples can be imaged in liquid or air. Fluorescence labeling is a well-developed technique that can be used to localize molecular components.
Confocal microscopy	Micrometer level	Confocal scanning microscopy enables three-dimensional studies of biological objects. Resolution techniques that break the optical resolution barrier are becoming available.
Field emission scanning electron microscopy (FE-SEM)	Nanometer level	For FE-SEM imaging, the sample is placed in a vacuum. Sample coating may be needed, as the technique generally requires an electron-conductive sample. The electron beam is used to probe the surface, and techniques for heavy metal labeling of surface molecules are often used.
Transmission electron microscopy (TEM)	Nanometer level	Image contrast depends on impeding electrons as they pass through the sample, usually by heavy metal staining. Operates under vacuum with resolution depending primarily on image contrast through staining. New advances allow imaging samples in a liquid cell.
Scanning tunneling microscopy (STM)	Nanometer level	Allows a relatively flat surface to be imaged by rastering a biased-atomically sharp needle point over a conducting (or semiconducting) surface. Samples can be imaged in ambient conditions and inside various electrolytes. STM can provide images down to atomic and molecular resolution as well as provide 3-dimensional visualization of the surface. Atomic manipulation of atoms and molecules can be achieved with an STM to create novel nanostructures.
Atomic force microscopy (AFM)	Nanometer level	Imaging is accomplished by monitoring the position of a sharpened tip attached to a microcantilever as it is scanned over a sample surface. Samples can be imaged in liquid or air with nanometer resolution at atmospheric pressure enabling dynamic studies. AFM provides three-dimensional surface visualization and measurement of nanomechanical properties of the sample.

direct methods of gaining data via images. Alternatively, indirect methods such as X-ray diffraction (XRD), ultraviolet-visible (UV-Vis) spectroscopy, thin-layer chromatography (TLC), Raman spectroscopy, and dynamic light scattering (DLS) provide important information on the composition, structure, and properties of nanometer-scale materials. The following sections give a brief description of microscopy techniques and characterization techniques that can be used in the various laboratory exercises listed in Chapter 5 and the projects listed in Chapter 6.

3.2 SCANNING ELECTRON MICROSCOPY

The SEM is an instrument that uses electrons instead of photons of light to form high-resolution images (micrographs). Its components are similar to the optical microscope, but instead of glass lenses, electromagnetic lenses are used to focus the electron beam onto the sample's surface. Unlike the optical microscope, which illuminates the whole sample, the electron beam scans very small areas of the sample at a time to create the image. The SEM has several advantages over conventional light microscopes. For example, images generated by the SEM have a large depth of field that allows substantial parts of the specimen to remain in focus compared to an optical microscope, for which only the focal plane is in sharp focus. Also, the SEM has a much higher resolution and can even resolve features down to about 2 nm in sophisticated machines such as the field emission scanning electron microscope, (FE-SEM). Furthermore, the fact that the lenses are electromagnetic gives the operator greater control to enhance the magnification range further. For example, the SEM magnification range can be adjusted from ×10 to ×100,000. Another advantage of the SEM arises from the interaction between the electron beam and the sample. When the instrument is set in backscatter mode, the interaction can be used to identify the elemental composition of the sample. For this reason, the SEM backscattered mode is routinely used by geologists to identify the elemental composition of rock and mineral samples. These features permit the SEM to produce striking images with great clarity, and it is for this reason that the SEM is one of the most useful scientific instruments available today.

The SEM, unlike light microscopy, must operate in the extremely low pressures of a vacuum chamber to sustain high-voltage electron beams. The beam voltage of most commercially available SEMs can range from 2 to 40 kV. The source of electrons used in the SEM beam comes from thermionic emission from a tungsten filament or a lanthanum hexaboride (LaB_6) cathode. These electrons are first collimated and then focused onto the sample as a fine e-beam by a series of electromagnetic lenses as shown in Figure 3.1a.

The electron beam interacts strongly with the electrons of the surface atoms in the sample and has the potential to interact in a number of ways. For example, the incoming electron can be scattered once or multiple times. During this scattering, the electron collisions may be elastic or inelastic in

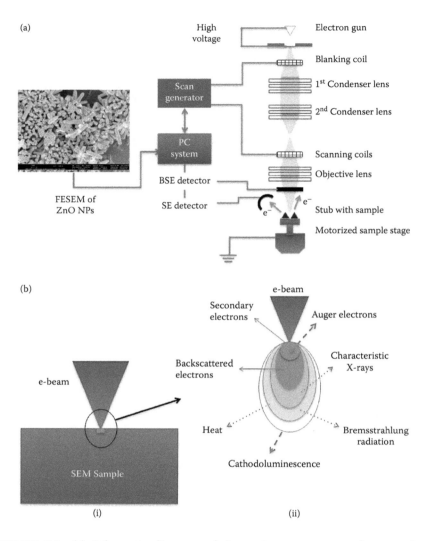

FIGURE 3.1   (a) Schematic diagram of the main components of a scanning electron microscope and the kind of image that can be acquired by this instrument. (b) (i) Specimen–electron beam interaction volume and (ii) e-beam interactions with the atoms of the specimen.

nature, and the scattering probability is dependent on the scattering cross section and the mean free path of the electrons. These parameters are also dependent on the size of the sample and the properties of the sample. When the electron beam is focused onto the sample, the resulting illuminated pear-shaped region, known as the interaction volume, penetrates into the surface as seen in Figure 3.1b(i). The depth of penetration ranges from 1 to 5 μm and is dependent on the beam voltage and the density of the sample. The beam-surface interactions generate important topographical information and data regarding the material's properties.

During the beam-surface interaction, several mechanisms take place, such as those involving secondary electrons (SEs), backscattered electrons (BSEs), and X-rays, as shown in Figure 3.1b(ii). SEs are ejected from the sample after colliding with atoms in the upper layers of the surface and are collected by the SE detector and used for imaging the surface topography. In addition to the emission of SEs are the BSEs, which are detected using a solid-state detector (Everhart-Thornley detector). The intensity of BSE is dependent on the atomic number of the sample material, accelerating voltage, and interaction volume. Importantly, the BSEs are used to create compositional maps of the specimen. However, because the interaction volume is greater for BSEs, the resolution is usually of lower quality than those of SEs. In addition to the SEs and BSEs, the SEM also produces characteristic X-rays, which can be used to determine the elements present in the sample material. Therefore, by scanning the sample surface topographical features and composition, data can be viewed, stored, or recorded as a micrograph.

Additional features of the SEM include a sample holder tilting function and selection of detector position, both of which can be used to produce micrographs with crisp, clear images with good contrast for enhanced visual presentation. Table 3.2 summarizes the types of information that can be gained from typical SEM investigation; Laboratory 5.11 gives the opportunity to carry out your own nanosample analysis using the

TABLE 3.2   Summary of the Types of Information from an SEM Investigation

E-Beam Interaction	Information	
Secondary electrons	Size, shape, and morphology of specimen at high resolution	Three-dimensional images are also possible.
Backscattered electrons	Atomic composition/ crystallographic composition and orientation by electron diffraction	Compositional maps are possible as well.
X-rays	Identification of sample's elemental components	
Cathodoluminescence	Fluorescence and optoelectronic properties	

SEM technique or using it to discover some of nanofeatures of natural nanomaterials, such as the structures on a butterfly wing or the highly regular nanoplates that form an oyster shell.

3.3 TRANSMISSION ELECTRON MICROSCOPY

In 1931, Knoll and Ruska developed the TEM; this was followed 8 years later by the first commercially available instrument. These early instruments were capable of producing high-resolution images that were superior to those of the high-resolution light microscopes of the period. The operating principle of the TEM is similar to the SEM except that the detector is a phosphor screen or plate capable of capturing the image. In fact, a TEM uses its electron beam in much the same way as a conventional light projector. In the case of the TEM, energetic electrons from the source are accelerated as they pass through a set of condenser lenses toward the sample. The electrons then pass through the sample, which means that the sample being analyzed must be thin enough for electron transmission to take place. During their passage, the electrons are scattered and must be collected and then focused by a set of objective lenses. The electrons are then magnified by a set of magnifying lenses (projector lens) before being projected onto a phosphor screen as presented in Figure 3.2. Projecting the

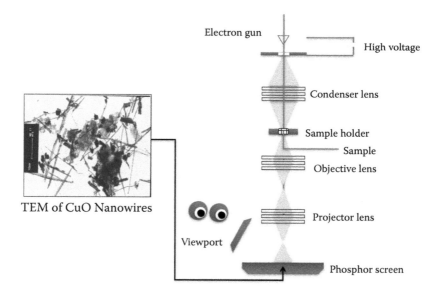

FIGURE 3.2 Schematic diagram of the main components of a transmission electron microscope and the kind of image that can be acquired by this instrument.

electrons to the screen requires large accelerating voltages around 300 kV, which is much higher than the 50 kV for a conventional SEM. The wavelength of the electrons projected at the screen is around 4 to 5 pm, and this is the reason why the TEM can provide such high resolution. Typically, the TEM can resolve features around 0.2 nm, which is close to the atomic radius of some materials.

Because the transmission of the electrons through the sample is inversely related to the thickness of the sample, materials with low atomic weight will transmit more electrons than those of materials with higher atomic weight. For this reason, the low atomic weight materials appear light on the screen, while the high atomic weight materials will appear darker. In addition, as the electrons pass through the sample, they can be diffracted between the atomic centers and thus produce electron diffraction patterns. This phenomenon is then used as a tool for determining the atomic spacing in the sample. And, because the atomic spacing is inherently different for each material, the TEM can be used to identify the structure of the sample, in particular the crystal structure and phase. Furthermore, biologists have used the TEM for studying infection by pathogens such as viruses and prions and to decipher the substructures of cellular materials.

3.4 SCANNING TUNNELING MICROSCOPY

In 1981, Gerd Binnig and Heinrich Rohrer (then at IBM, Zurich) invented the first STM for imaging surfaces at the atomic level [1]. This was 10 years after a similar instrument called the surface profilometer (or topografiner) was developed by Young et al. (at the National Bureau of Standards) [2]. Because the high resolution of the STM (~0.1-nm lateral direction), it has been used to study structures such as single-wall carbon nanotubes and atomic silicon [3,4]. The operational principle of STMs is based on the phenomenon of quantum tunneling. In this mechanism, a conducting STM tip is lowered close to the sample's surface, and then a voltage difference (bias) is applied between the tip and the surface. This results in electrons tunneling through the vacuum to produce a tunneling current flow through the tip and the surface. The tunneling current is dependent on the height of the tip above the surface, the applied voltage, and the local density of states (DoS) in the surface of the sample. This extremely small current flow (about the order of nanoamperes) is measured to provide information on the sample's surface. In a vacuum, the tunneling current can be approximated using Equation 3.1, where k is a constant dependent on the work function of the surface and the tip, and d is the distance between the surface and the tip.

$$I_t \propto e^{-kd} \tag{3.1}$$

There are generally two STM operational modes; the first is constant height, and the second is constant current. In a typical STM investigation, the sample is slowly moved toward the tip or vice versa depending on the manufacturer's model. During this adjustment, an optical microscope can be used to preposition the tip a few micrometers away from the sample. Then, a drive assembly controlled by a personal computer is engaged to move the tip closer to the sample's surface and at the same time test for a tunneling current. This procedure is continued until a tunneling current is detected; once detected, the STM is ready to scan the sample. This mode is called the constant-current mode because the tip is moved up and down via an electronic feedback loop to maintain a constant tunneling current as seen in Figure 3.3. Generally, an STM can scan an area 20×20 μm^2 (x–y plane) at a rate of 1 Hz, and a scan rate of 10 Hz usually can be used for small samples. A smaller scan gives higher resolution, which makes it possible to see more morphological features on the surface of the sample.

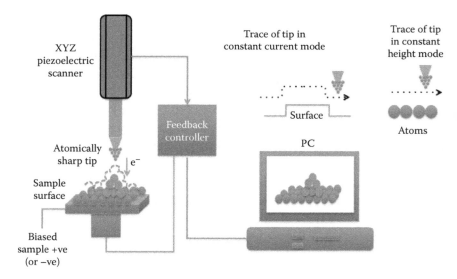

FIGURE 3.3 Schematic diagram of a scanning tunneling microscope. The essential components of the tip, piezoscanner, and controller/PC are shown. The modes of operation are also shown.

In the second mode, the tip height above the sample is kept constant, and fluctuations in the tunneling current are monitored. During this mode, the STM gathers information on the local DoS and the atomic or molecular nature of the sample. Thus, atomic and molecular resolution of a surface is possible in the constant-height mode. The STM not only has the ability to operate in the ultravacuum range but also to operate in air, liquids, and gases. Another interesting feature of the STM is the ability to use the tip to create local atomic features. For example, Eigler and his team (at IBM Almaden) using a tungsten (W) tip in a low-temperature ultra-high vacuum were able to move Xe atoms at will over a crystal surface of nickel (Ni) (110) [5]. More impressively, in 2012, IBM researchers announced that they could store a byte of information using just 12 atoms.

3.5 ATOMIC FORCE MICROSCOPY

Today, the AFM is one of the most powerful tools available to nanoscien-tists and nanotechnologists for examination of surfaces, imaging atoms or molecules, and manipulating these atoms and molecules. The technique does not require high-vacuum environments like other techniques and can even be operated with samples within its own liquid environments. The basic principle behind the operation of an AFM is straightforward, but it does need sophisticated instrumentation and vibration control to achieve high resolution. The AFM consists of an extremely small flexible beam with a tip at one end. When the tip is brought into close contact with the surface, it will be deflected by the sample surface's forces. Once stable, the tip is scanned (rastered) across the sample. The tiny deflections are monitored with a sensor system (usually a laser and a position-sensitive detector-photodiode), and the tip-sample distance is kept constant by a feedback mechanism in a similar way as its predecessor, the STM. The data gathered provide a direct map of the various forces occurring between the tip sensor and the surface. The magnitude of the force recorded is usually on the order of nanonewtons. The tip is indeed very sharp, terminated the end with a radius of around 50–20 nm. An example of a typical AFM system is shown in Figure 3.4a.

Because the AFM senses forces, it can be used to map the surface of all materials, such as metals, insulators, and semiconductors. The forces encountered by the AFM tip can either be attractive (van der Waals) forces or repulsive electrostatic forces. The net force exerted on the tip is a function of the distance between the tip and the sample. Figure 3.4b presents a typical AFM tip being lowered onto a sample surface and the forces encountered by the tip. The AFM's theoretical magnification can be up to 10^9 and can routinely image atoms and molecules. During the development of the technique, several operational modes were found; however, only three main modes of operation are generally used, namely, contact, noncontact, and tapping. These modes are detailed in Table 3.3.

Different cantilever-tip combinations are now available; the cantilevers are fabricated from a single crystal of silicon. The length of AFM contact and tapping mode cantilevers is between 100 and 500 μm, the width can

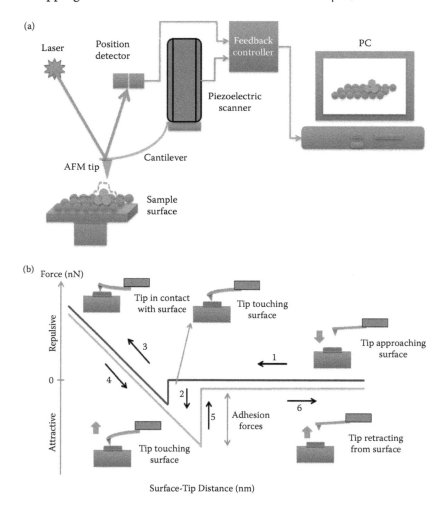

FIGURE 3.4 (a) Schematic diagram of an atomic force microscope (AFM) and its essential components. (b) Force curve of a cantilevered tip approaching and retracting from the surface to be imaged in an AFM.

(c)

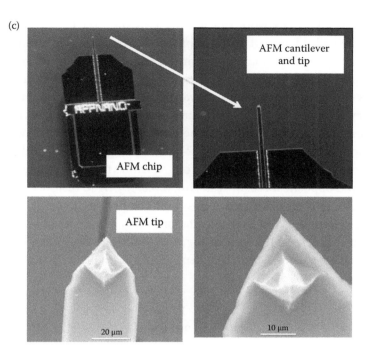

FIGURE 3.4 (*Continued*)   (c) An Si-based AFM tip with typical dimensions.

TABLE 3.3   AFM Modes of Operation and Capabilities

AFM Mode	Configuration	Capability
Contact mode	Repulsive mode Static	Hard materials
Tapping mode	Oscillatory mode, either frequency or amplitude	Biological materials such as DNA and polymer work
Noncontact mode	Weak-attractive regime Oscillating probe can be applied with water layer	Soft surfaces

range from 25 to 40 μm, and the thickness can vary between 1 and 10 μm. The resonant frequency of these cantilevers ranges from 10 to 300 kHz, and the spring constant varies between 0.1 and 50 Nm^{-1}. The tip diameter ranges from a few nanometers for finer resolution to 20 nm for a standard AFM image (largest is around 50 nm). Noncontact AFM mode cantilevers are made of a highly doped single crystal of silicon. In the early days of the AFM, research groups generally handcrafted their own tips using a variety of materials. An example of a silicon-based AFM tip is shown in Figure 3.4c.

For the AFM to work efficiently, the cantilever-tip assembly must be specifically designed to have properties that can withstand the tip forces and bending moments encountered during the scanning process. Generally, when a material is acted on by a force F, the atoms or molecules of the material will move away from their equilibrium position and will try to return to their initial state of equilibrium. This is true for an elastic spring system, which can be modeled using Hooke's law. The law is expressed by $F = -k\Delta x$, where k is the spring constant, and Δx is the displacement or deflection. And, from Newton's third law, there will be an equal and opposite force (or restoring force) trying to restore the atoms or molecules to their equilibrium position. The potential energy of an ideal spring system is given by the equation $PE = \frac{1}{2}kx^2$. So, when a mass m (loaded mass) is added to a spring, the resulting system can oscillate at a natural frequency f, which is dependent on k and m by the following relation:

$$f = \frac{1}{2\pi}\sqrt{\frac{k}{m}} \tag{3.2}$$

In an AFM, the cantilever-tip assembly acts like a harmonic oscillator with a natural frequency determined by the dimensions of the cantilever. To operate effectively, the cantilever-tip assembly must be sensitive for a given force as well as stiff enough to cope with the high resonant frequency and deal with any environmental noise vibrations. Therefore, k needs to have a low value, while the f is high. This is achieved by having a high k/m ratio (R). Consequently, an AFM tip with a low spring constant (<1 Nm^{-1}) together with a high resonant frequency (>10 kHz) is generally preferred for AFM microscopy.

The AFM field has benefited significantly from production technologies used in manufacturing silicon-based electronics and PCs in the late 1980s. These technologies were able to make reproducibly sharp tips, and the batch fabrication techniques used resulted in the cost of micromachined cantilever-tip assemblies becoming less expensive. In addition, the mechanical properties of the cantilever-tip assembly could be exactly controlled to produce high levels of reproducibility in the AFM operations. AFM cantilever-tip assemblies are usually fabricated from solid silicon and silicon nitride (Si_3Ni_4) because these materials have been extensively studied and used in the PC industry for several decades. Typically, a cantilever-tip structure is undercut and released from the parent material by either a wet or dry etching technique. The deflection of the cantilever is calculated from Equation 3.3:

$$\delta = 3\sigma (1 - v)/E(L/t)^2 \tag{3.3}$$

where v is the Poisson ratio of the material, E is Young's modulus of elasticity, σ is the applied stress, and δ is the resulting deflection.

The deflection δ is related to the spring constant via $k = F/\delta$ and can be expressed as

$$k = Ewt^3/4L^3 \tag{3.4}$$

where L is the length of the cantilever, and t is the cantilever thickness. A typical resonant frequency of an AFM tip is around 400 kHz, with a force constant of 0.8 N/m. The advantage of using precisely machined monoxtal silicon or silicon nitride for the cantilever-tip assemblies means both the resonance frequency and the force constant can also be calculated accurately; hence, no calibration is required. Currently, there are several commercial cantilever-tip manufacturers that can supply assemblies with well-defined characteristics to fit a wide range of AFM units and applications. In Laboratory 5.12, you will have the opportunity to examine your own synthesized nanomaterial samples with AFM.

3.6 X-RAY DIFFRACTION

X-ray diffraction is a versatile nondestructive method used in material, physical, chemical, and biological sciences to identify and quantitatively determine the crystalline phases present in a solid sample. Powder XRD is a technique specifically developed to characterize the crystallographic structure, crystallite size (grain size), and preferred crystal orientations in powdered solid samples. In X-ray crystallography, a collimated beam of X-rays is generated and sent toward the sample material under investigation. The atomic centers in the crystal lattice act as point diffractors when the X-ray beam illuminates the sample material.

Each of the diffracted X-ray beams corresponds to a coherent reflection (called a Bragg reflection) created by the atomic planes containing the atoms with the crystal. In fact, X-rays and the Bragg equation have been extensively used in identifying crystal structures (and the atomic arrangements) since the early 1900s. When an X-ray source of known wavelength λ

is sent through an unknown crystal structure (or crystals), the diffracted beam angle is predicted by the Bragg equation.

$$n\lambda = 2d \sin \theta \qquad (3.5)$$

where n is any integer, λ is the wavelength of the incident X-rays, d is the interplanar spacing, and θ is the diffraction angle. Figure 3.5a presents the atomic planes and reflection angles of the X-ray beams from two

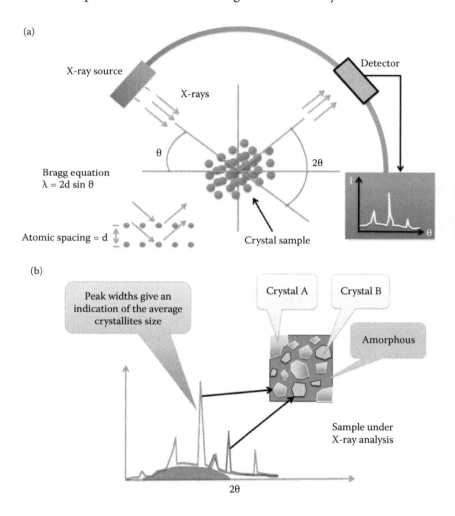

FIGURE 3.5 (a) Determination of crystalline structure by the X-ray diffraction (XRD) technique. (b) XRD plot of a typical sample with more than one crystalline phase.

(c)

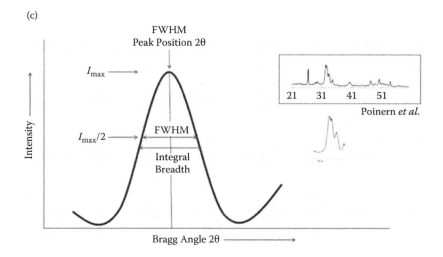

FIGURE 3.5 (*Continued*)   (c) (i) Peak width-full width at half maximum and (ii) a typical scan of nanohydroxyapatite XRD plot.

successive planes. A typical instrumental setup is also shown Figure 3.5a, and the results of an X-ray study of a sample are usually presented as a plot of intensity versus 2θ. The three indices *h*, *k*, and *l* are used to define each plane, which in turn relates to the atomic spacing *d*. In the simplest form, the relation between *d* and *hkl* in a cubic structure (such as NaCl) are presented in Equation 3.6:

$$d = \frac{a}{\sqrt{h^2 + k^2 + l^2}} \tag{3.6}$$

where *a* is the lattice spacing of the cubic crystal, and *h*, *k*, and *l* are the Miller indices of the Bragg plane. Figure 3.5a also illustrates how a crystalline structure (shown diagrammatically as a latticework of atoms) may be determined through XRD. As the crystal and detector rotate, X-rays diffract at specific angles, and the detector measures the intensity (*I*) of X-rays. Angles of diffraction (where the Bragg equation is satisfied) are marked by peaks, and the peak height is a function of the interaction between the X-rays and the crystal and the intensity of the source. In terms of the Bragg equation, we are looking for *d* as we change the angle θ and the wavelength is kept constant.

A crystal is an ordered array of atoms of molecules arranged in a definite three-dimensional structure. As mentioned, above common salt (NaCl) is a typical cubic structure. Atoms of Na^+ ions and Cl^- ions are arranged in a regular pattern within the crystal lattice, with an Na^+ ion surrounded by six Cl^- ions. This periodic structure is distributed throughout the whole solid and gives rise to parallel planes of atoms with an interplanar spacing of d. The spacing is highly dependent on the nature of the crystal, and this can be used to identify this particular material phase. In many cases, a material contains more than one type of crystal, and XRD can be used to identify the various types of crystals inside the material as well as show any amorphous fraction (see Figure 3.5b).

A typical diffraction spectrum consists of a plot of reflected intensities versus the detected angle 2θ. The 2θ values of the peak depend on the wavelength of the anode material of the X-ray tube. The wavelength of X-rays is known because the anode material and the energy of accelerated electrons are known. The single-wavelength X-rays used in diffraction studies are K_α radiation. Extracting information from the X-ray spectrum is done by comparing the diffraction data of the unknown sample with recorded diffraction data in the International Center for Diffraction Data (ICDD) data base. The average particle size D of the sample can be estimated from XRD data using Scherrer's formula:

$$D = \frac{K\lambda}{FWHM \, \cos\theta} \tag{3.7}$$

where K is the Scherrer constant, $FWHM$ is the full width at half-maximum value (in degrees) of the reflection peak (see Figure 3.5c(i)) that has the same maximum intensity in the diffraction pattern, λ is the wavelength of X-rays, and θ is the diffraction angle of X-rays. The Scherrer constant K in the formula accounts for the shape of the particle and is generally taken to have the value 0.9. The size obtained from the Scherrer formula yields the apparent or average particle size of the sample. Powders of materials are generally aggregates of smaller particles and thus consist of a distribution of various particle sizes. An example of an XRD pattern of a nanocrystalline hydroxyapatite (HAP) powder is shown in Figure 3.5c (ii). In this case, the nanometer-scale HAP was formed by first creating calcium triphosphate nuclei by ultrasound in aqueous conditions followed by thermal treatment to produce the nano-HAP.

3.7 UV-VIS SPECTROSCOPY

Spectroscopy is the study of interactions occurring between radiation and matter. It started in 1665 with white or visible light being made to split into its individual components by a simple glass prism. This is also the reason why the colors of the rainbow appear in the atmosphere during rainy days. The atoms or molecules of many materials have properties directly related to color and the intensity of the radiation being absorbed or generated. For example, when energy in the form of UV or visible light illuminates certain molecules in a solution containing π electrons or non-bonding electrons, the absorbed energy from the light will excite the electrons and lift them to higher antibonding orbitals, which are normally empty. UV-Vis spectroscopy is reliant on this type of absorption phenomena, and the technique is used to study the transition of metal ions, highly conjugated organic compounds, and metallic nanoparticles such as those found in gold colloids. Studies have found that there is a direct relationship between the concentration of the analyte (molecule of interest) and the path length through which the light passes through the solution and the amount of absorption. The relation is expressed as Beer Lambert's law and is stated as

$$A = \varepsilon l c \qquad (3.8)$$

where A is the absorbance of the sample, ε is the molar absorptivity or molar absorption coefficient, l is the length of the light path in the solution (cm), and c is the concentration of the solution (moles/L).

In a UV-Vis spectroscope, light is generated in the visible and UV range by a lamp (tungsten, deuterium, or xenon) and then split into different wavelengths by a diffraction grating. The different wavelengths of light are then projected through the sample material of interest, and the resulting light is detected. The absorbance or transmission profile of the sample material is then plotted as intensity versus wavelength. Generally, the UV range is between 190 and 380 nm, and the visible range is between 380 and about 750 nm. Both these types of radiation interact with matter and cause electronic transitions (promotion of electrons from the ground state to a high-energy state). Figure 3.6a presents six electronic transitions that are possible in these regions. They consist of π to π*, σ to σ*, σ to π*, π to σ*, nonbonding to σ*, and nonbonding to π*. When sample molecules are exposed to light in the UV-Vis region, some of this light energy will be absorbed by the electrons in these molecules and be elevated into

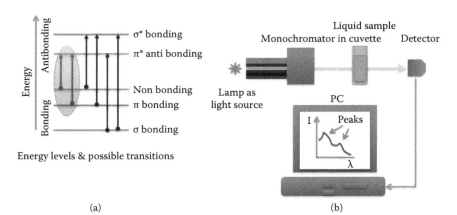

(a)   (b)

FIGURE 3.6   Schematic of ultraviolet-visible (UV-Vis) spectrometer, electronic transitions, and the associated instrumentation.

higher-energy orbitals, so the π-to-π* and nonbonding-to-π* transitions are observed. The spectrophotometer will record the absorption at each wavelength, and then the resulting spectrum is plotted as shown in Figure 3.6b.

3.8 THIN-LAYER CHROMATOGRAPHY

Today, the term *chromatography* relates to a family of separation techniques relying on the distribution of different materials between two phases, namely, a stationary phase and a mobile phase. Generally, it provides the scientist with qualitative results quickly and is considered the first pass for identification of pure compounds or molecules from a mixture of compounds and isolating components of mixtures or synthesized compounds. For example, the pharmaceutical industry requires bioactive compounds with very high purity levels, and forensics science needs to differentiate between several DNA sources to match a perpetrator to a certain crime scene or event. In these cases, chromatography plays an important role in the identification and isolation of pure extracts. Other more quantitative techniques such as gas chromatography (GC) and high-pressure liquid chromatography (HPLC) can then be used for the subsequent steps. A common chromatographic method used in chemistry and biological sciences for separating mixtures is TLC. This method is quick, sensitive, and above all simple to use in the laboratory. It can be readily used to check the purity of a compound and to monitor the progress of a reaction currently under way.

A typical TLC plate consists of a thin silica plate (~200 μm thick) and is used as the stationary phase. The thin layer of silica gel is deposited onto a glass plate or thick aluminum foil with a binding agent. The TLC technique is similar to paper chromatography, and both techniques are available in a variety of commercial products and compound identification procedures. In the TLC method, compounds are carried along in a mobile phase; along the plate, various components separate and deposit (fraction spot) as shown in Figure 3.7. The components can be identified by their R_f (retention factor) value. This value is qualitative because many variables can affect it. If the experimental conditions of the TLC are exactly reproduced and strictly controlled, then the R_f value can be considered an absolute value.

In practice, a fraction (spot) of an unknown compound together with other standard or control (pure) compounds suspected of being in the unknown compound are placed on one end of the TLC plate, and the chromatographic run is started. As the mobile phase moves across the TLC plate, components in the unknown compounds slowly separate out. As the R_f value of identical compounds will be the same under the same conditions, components can be identified by direct comparison with the standard spots, which also separate out on the TLC plate.

$$Rf = \frac{distance\ moved\ by\ component}{distance\ moved\ by\ solvent} \qquad (3.9)$$

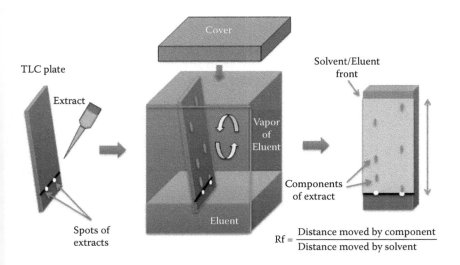

FIGURE 3.7 Schematic diagram of the thin-layer chromatographic (TLC) technique to separate and identify the components of an extract.

Sometimes, components may be invisible to the naked eye, so small amounts of fluorescent material are sprayed onto the plates, which are then illuminated with UV light to assist in identifying these components. Laboratory 5.4 will give you the opportunity to use the TLC technique in determining fractions from carbon soot generated by a combustion process.

3.9 RAMAN SPECTROSCOPY

Raman spectroscopy is named after Sir Chandrasekar Raman's discovery of a particular type of light-matter interaction in 1928. When light interacts with matter, it is scattered in three modes that can be observed. The first is Rayleigh scattering, the second is Stokes-Raman scattering, and the third is anti-Stokes-Raman scattering. In Rayleigh scattering, there is elastic scattering of light by matter particles much smaller than the incident radiation; it was originally discovered by Lord Rayleigh in the 1860s. This phenomenon explains the blue color of the atmosphere and is caused by the scattering of sunlight by air molecules. This is the predominant scattering mode. However, there is also the possibility that the incident photons of light could interact with the molecules of the specimen in such a way that energy is either gained or lost. This type of interaction is inelastic and is called the Raman effect. After this molecular interaction, involving the chemical bonds of the molecule, there is a shift in the frequency of the light resulting from the release of a photon of light. Depending on the vibrational energy states of the molecules, the Raman photons can be shifted to either higher or lower energy levels. Therefore, by analyzing the change in frequency of the original photon, the types of molecules present in the sample can be determined because specific molecules will absorb specific amounts of energy. Furthermore, any strain acting on the molecules will also be reflected in subsequent changes in frequency. Despite the Raman process being relatively weak compared to Rayleigh scattering, it can still provide significant information regarding the vibrational modes of molecules.

Today, Raman spectroscopy has been used to study solid, liquid, and gaseous samples and has several advantages over other analytical techniques. A typical experimental setup is presented in Figure 3.8 and shows a small fraction of inelastically scattered Raman radiation from the sample first directed to a mirror, then directed through a grating, and finally directed to a detector. The detector interfaces with a PC equipped with

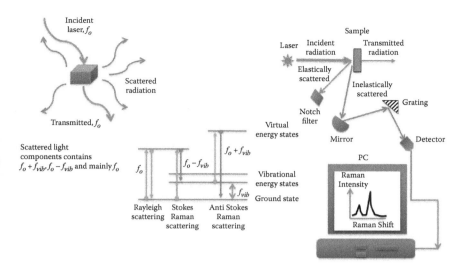

FIGURE 3.8   Schematic of the energy transitions in Raman spectroscopy and the associated instrumentation.

an interface and software that generates a plot of intensity against Raman shift (in cm^{-1}).

Raman spectroscopy has been used to study surfaces undergoing corrosion, electrochemistry, and catalysis. It has also been used to detect and identify absorbed molecules without the need for techniques based on ultrahigh vacuum (UHV), which makes Raman spectroscopy a straightforward and easy technique to use. Furthermore, extreme enhanced sensitivity is also observed when various chemical species are deposited onto nanogold and nanosilver. This enhancement is reliant on the surface plasmon resonance (SPR) of the respective nanometer-scale metal and has found potential applications in biological sensors. The SPR enhancement phenomenon has not been fully elucidated to date, but it has already been used in the study of biological moieties of DNA and peptides. Enhancement of the vibrational signal of some chemical species absorbed onto nanosilver have been measured on the order of 10^6, and single molecular detection is also possible using surface-enhanced Raman scattering (SERS) [6].

3.10 DYNAMIC LIGHT SCATTERING

Nanotechnology and nanoscience are strongly reliant on the fact that the properties of nanometer-scale materials have a direct relationship to their size and shape. Therefore, the size has a large bearing on the final characteristics of the nanometer-scale material being synthesized. Generally, particles that are synthesized in a liquid can be quantified using a straightforward technique known as dynamic light scattering (DLS). This technique is also called photon correlation spectroscopy (PCS) or quasi-elastic light scattering (QELS). The technique is noninvasive and can measure the size and distribution of particles in the micrometer to the submicrometer range. Recent technological developments have enabled this technique to measure particles in the nanometer range and also to help determine the overall morphology of the particles in solution.

Particles suspended in solution or dispersed in a solvent are always in constant motion because of Brownian motion. Random collisions between solvent molecules and particles will also affect this motion. Therefore, small particles will move faster compared to larger (heavier) particles. When a beam of light passes through a colloidal solution, the particles will scatter light in all directions. If the particles are very small compared to the wavelength of light, then the intensity of the scattered light will be uniform in all directions (Rayleigh scattering). However, when the particles are larger than 250 nm in diameter, the intensity of the scattered light is dependent on the scatter angle (Mie scattering). If a monochromatic light in the form of a laser is used, then it is possible to record time-dependent fluctuations in the scattered intensity using a photomultiplier detector. The fluctuations are caused by the Brownian motion of the particles within the illuminated zone and hence contain information about their motion and the velocities of the particles. The random particle motion and particle size analysis can be modeled using the Stokes-Einstein equation (Equation 3.10). The equation directly relates the diffusion coefficient measured by dynamic light scattering to particle size,

$$D_h = \frac{k_b T}{3\pi\eta D_t} \tag{3.10}$$

where D_h is the hydrodynamic diameter of the particle, D_t is the translational diffusion coefficient (from DLS measurements), and k_B is the dynamic viscosity. The fully automated technique is repeatable, has fast acquisition times (within a few minutes), and measures particle sizes and mean size distributions. Moreover, the concentration of the sample is not a factor, and even turbid samples can be processed. Figure 3.9 presents a schematic diagram of a typical DLS system used to measure particle size.

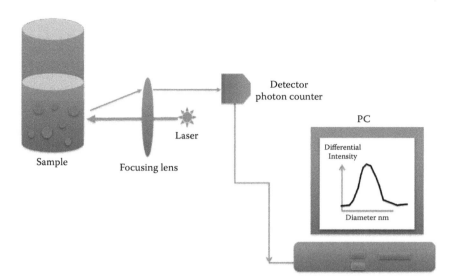

FIGURE 3.9   Schematic diagram of a dynamic light-scattering system used to measure particle size and the associated instrumentation.

Another material property that can be measured using a DLS system is the zeta potential of the particle, which represents the overall charge on that particle. The magnitude of the zeta potential represents the magnitude of the repulsive force produced by the particle and gives an indication of the stability of the particle in solution. For example, nanogold has a low zeta potential, which equates to low repulsive forces in solution and results in particle aggregation. This measurement can assist in generating stable colloidal solutions.

REFERENCES

1. Binnig, G.; Rohrer, H.; Gerber, Ch.; Weibel, E. Tunneling through a controllable vacuum gap. *J. Appl. Phys.* **1982**, *40*, 178.
2. Young, R.; Ward, J.; Scire, F. The topografiner: an instrument for measuring surface microtopography. *Rev. Sci. Inst.* **1972**, *43* (7), 999–1011.
3. Venema, L.; Wildöer, J.; Dekker, C.; Rinzler, G.; Smalley, R. STM atomic resolution images of single-wall carbon nanotubes. *Appl. Phys. A* **1998**, *66*, S153–S155.
4. Binnig, G.; Rohrer, H.; Gerber, C.; Weibel, E. 7 x 7 reconstruction on Si/111/ resolved in real space. *Phys. Rev. Lett.* **1983**, *50*, 120–123.
5. Stroscio, J.A.; Eigler, D.M. Atomic and molecular manipulation with the scanning tunnelling microscope. *Science* **1991**, *29*, 1319–1326.
6. Kneipp, K.; Wang, Y.; Kneipp, H.; Perelman, L. T.; Itzkan, I.; Dasari, R. R.; Feld, M. S. Single molecule detection using surface-enhanced Raman scattering (SERS). *Phys. Rev. Lett.* **1997**, *78* (9), 1667–1670.

Laboratory Safety and Scientific Report Writing

4.1 CHAPTER OVERVIEW

This chapter has two sections. The first section prepares the student for working in a laboratory setting. This begins by discussing the importance of good laboratory practices and work ethics; this is followed by hazard identification and management and concludes by discussing safety procedures that should be followed in a teaching laboratory. The second section deals with scientific report writing, which is an essential part of undertaking any scientific research. The results of any experimental work carried out in the laboratory must be recorded in a logical fashion so that all the data resulting from the procedure used are collected. This data then need to be translated into a well-written report so that other researchers can readily understand the material presented. Remember that there is no point carrying out a research project if nobody can understand what you have written. Therefore, proper scientific report writing is an important factor in conveying experimental results and promoting the advancement of science. A typical report format is presented and explained; this is designed to assist students write the results of their experimental work carried out in Chapter 5.

4.2 INTRODUCTION TO LABORATORY SAFETY

The laboratory is a safe place only when you follow the safety rules. The laboratory can be an extremely dangerous place to be if you are unfamiliar with safe procedures for handling chemicals, operating apparatus and instruments, and performing experimental work. A laboratory is a place with many potential hazards that can result in serious injury or even death if you are not cautious and you do not follow safety procedures. In particular, hazards such as improper operation of laboratory equipment and conducting experimental work by inexperienced personnel and students can result in injuries and even death to others. Laboratory hazards are unavoidable because many of the materials used are potentially toxic. However, by following safety procedures and taking sensible precautions, these potential hazards can be significantly reduced to manageable levels. Only when working in a safe environment and following all the laboratory safety procedures can a student gain sufficient experience and expertise in handling and working with chemicals [1].

4.2.1 Good Laboratory Practices

Good laboratory housekeeping comes from experience, but the following discussion is a helpful start. To begin, it is **absolutely forbidden** to smoke, eat, or drink in the laboratory. As a student, you should not be working unsupervised; this is because there are many potential safety hazards that you may not be aware of or trained to handle. When you attend the laboratory class, a trained supervisor or laboratory attendant will assist you, so if you have any concerns or questions, ask this person. While every effort is made to reduce the risk of accidents, from time to time accidents do occur in the laboratory. In the event of an accident, summon the laboratory supervisor immediately and inform the supervisor of the incident. If you are dealing with the incident and are incapacitated, please ensure that someone else informs the supervisor. Do not be afraid to report an accident or injury, however minor it may be. Once notified, the supervisor will coordinate any action that is required to resolve the situation. To minimize accidents and warn personnel about potential dangers, an internationally recognized set of standardized safety signs is used to communicate potential hazards. For example, Figures 4.1 illustrate some of the commonly encountered hazard-warning signs that are normally found in a laboratory.

4.2.2 Preparation

It is important that you familiarize yourself with all the safety procedures, equipment, and chemicals present in the laboratory. You also need to know how to respond to any potential accidents that could occur in the laboratory. Therefore, on your first day in the laboratory follow the instructions the laboratory supervisor or laboratory attendant will give you regarding the location of fire extinguishers, accidental chemical spillage kit, first-aid kit, and the closest telephone in case you need to call for assistance. In addition, it is highly recommended that you read all laboratory safety and hazard documentation prior to actually commencing any experimental work. In this way, you will be able to identify any hazardous chemicals that you may need to use in your experimental procedures.

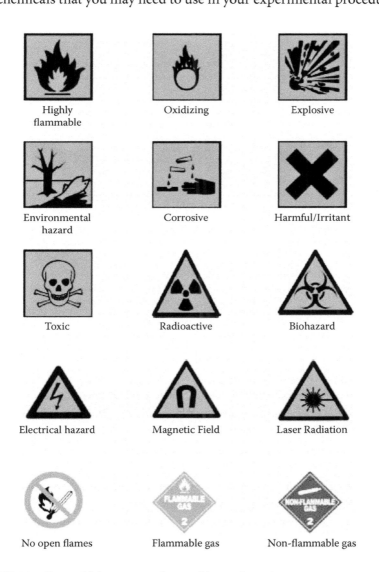

FIGURE 4.1   General laboratory safety and hazardous signs.

FIGURE 4.1 (*Continued*)   General laboratory safety and hazardous signs.

To assist you in hazard identification, every chemical in the laboratory will have a **Materials Safety Data Sheet (MSDS)**, which you can consult before starting your laboratory procedure. The data sheets contain detailed information on the chemical you are planning to use, the potential hazards the chemical has, and which procedures are needed to handle the chemical safely. Remember, by identifying the hazards associated with the chemical and following the handling procedure means that you are effectively reducing the risk of accident or mishap. Another important factor in reducing hazards is to always keep the laboratory and your work area clean and uncluttered during your laboratory sessions. At the end of each laboratory session, clean all glassware and laboratory equipment and store chemicals properly.

4.2.3 Protective Clothing

At all times, you must dress appropriately in the laboratory. It is not a place to wear expensive, fashionable clothing because unavoidable chemical spills can sometimes occur despite our best efforts at minimizing them. Short pants and skirts are forbidden, and to be admitted into the laboratory, your legs must be protected by wearing long trousers. Furthermore, a cotton laboratory coat must be worn at all times and will aid in reducing damage to your skin and clothes from any splashed or spilled chemicals. Moreover, the laboratory coat will reduce the risk of setting fire to any synthetic clothing materials you are wearing. Your feet must also be protected at all times by wearing shoes that completely cover your feet. Sandals and thongs are not acceptable in the laboratory. It is always a good idea to protect your hands from laboratory chemicals by wearing gloves. It is important to carefully select a pair of gloves that will protect you from the chemicals you will be handling. Laboratory gloves can be made from a variety of polymer-based and other materials, such as latex, butyl rubber, nitrile, natural rubber, and polyvinyl chloride (PVC). Also, gloves vary in thickness and provide different levels of protection against particular chemicals. Importantly, the glove manufacturer provides literature about the types of chemicals that can be used with the gloves and their performance. Remember, to reduce the risk of being injured, you must select a pair of gloves that are appropriate for the chemicals you will be using. If you have any doubts about selecting the right gloves, you should ask the laboratory supervisor or attendant, who will advise you of the appropriate glove type.

4.2.4 Eye Protection

Your eyes are the most precious sense you have, and you must protect them at all times. Whether you are mixing chemicals or just writing up your experimental results in your notebook, you must at all times wear safety glasses. Remember, no one is perfect, and another laboratory user might mishandle something and splash a chemical onto you; if you are not wearing safety glasses, the chemical might end up in your eyes. Therefore, to minimize the risk of eye damage, safety glasses must be worn in the laboratory at all times by everyone. There are no exceptions. Please note that prescription glasses do not provide adequate protection for your eyes in the laboratory because they were never designed to protect your eyes against splashed liquids. If you are required to wear prescription glasses, then you are required to wear safety glasses over them. Generally, contact lenses are not permitted in the laboratory because of the risk of chemicals being trapped between your eye and the lens, thus making it difficult to flush the chemicals away. If wearing contact lenses cannot be avoided, then you must wear well-fitting goggles to protect your eyes and inform the laboratory supervisor or attendant. People who do not wear safety glasses or well-fitting goggles will not be permitted to enter the laboratory. If some unforeseen accident occurs and chemicals do get into your eyes, swift action must be taken. You must alert the laboratory supervisor or attendant, who will get you to an eyewash station. At the station, the chemicals will be washed away from the affected eye or eyes using large quantities of water. If an eyewash station is not available, then someone will need to hold the injured person, and keeping the affected eye or eyes open, the eye or eyes can be flushed with large quantities of water to wash the chemical away. The incident must be reported to the laboratory supervisor or attendant who will obtain prompt medical assistance to ascertain if any eye damage has occurred.

4.2.5 Laboratory Hazards

4.2.5.1 Chemical Hazards

Chemicals are dangerous compounds that can be corrosive, toxic, oxidizing, flammable, or explosive in nature. Because there are many hazards associated with using chemicals, a special hazard labeling system is used to identify all the hazards associated with a particular chemical. This section gives some general procedures that should be followed when handling chemicals in the laboratory; Section 4.2.7 discusses the chemical labeling system.

Generally, chemical hazards in the laboratory can be reduced by dealing with the following hazards appropriately:

1. When using chemicals, never taste or sniff them; they can be extremely toxic.

2. Never carry chemicals around the laboratory in open containers and always use the proper receptacle for this purpose.

3. Always put your waste materials into specified areas denoted by your laboratory attendant and dispose of any toxic materials in the appropriate waste containers.

4. Good laboratory practice demands that all your glassware is properly labeled and dated.

5. Keep the exteriors of reagent bottles and stoppers clean and dry.

6. Never leave a stopper on the bench;, always replace the stopper and return the reagent bottle to its proper place immediately after use.

7. Do not return solutions or solids to a reagent bottle—it can only result in contamination. Transfer to clean, dry beakers or flasks only the minimum amount you need to perform the experiment.

8. When performing nanotechnology-based experiments, micrograms of contaminants can have a drastic effect on your experimental results. Never assume that a container or piece of equipment is clean until you have thoroughly cleaned it yourself.

9. All bottles containing chemicals or products should be clearly labeled to indicate what they contain. And, never combine chemicals randomly because you never know what chemical hazard you can create.

10. When making dilute acids and bases, always add the concentrated solution to water slowly while slowly stirring to solution; this will significantly reduce the risk of a violent reaction and prevent liquid sputtering back out of the container.

4.2.5.2 Glassware Hazards

Handling glassware is a routine task in any laboratory work, and care is needed to prevent breakages of this fragile material. Breakages can be expensive, and every effort must be made to reduce the risk of breakage. Remember that most breakages are caused by carelessness or not paying attention to what you are doing. Unfortunately, accidents do occur, and when they do, report the incident so that the broken glassware can be repaired or replaced, thus preventing inconvenience in the future. Glassware that cannot be repaired must be disposed of safely by placing the broken glass into special glass waste bins, which are provided in every laboratory. On no account is any broken glass to be put into other waste bins. If an injury occurs during the glass breakage, you must notify the laboratory supervisor or attendant immediately so that first aid can be applied. If you have any doubts about handling and using the glassware, it is best to ask the laboratory instructor; never rely on the accuracy of the advice of a fellow student. Remember that good experimental results will come from clean glassware and equipment. And, make it a habit to keep your bench space clean and dry at all times and to wash all dirty glassware before and after use.

4.2.5.3 Laser Light

In recent years, laser light has become a popular and useful tool in medical and scientific experimental procedures. Today, the laser is used in a wide variety of tools, ranging from bar code scanners found in most shops to more sophisticated scientific analysis equipment, such as that for Raman spectroscopy and dynamic light scattering (DLS), which need a coherent light source. In the laboratory, you will be using a simple pen laser light to indicate the presence of nanoparticles in solution. The most commonly used pen lasers are red (635 nm), green (532 nm), and blue (445 nm) and come in a number of power levels. Because pen lasers are small and maneuverable, they can be accidentally aimed into someone's eyes. Shining laser light into someone's eyes is extremely hazardous and should be avoided at all times because serious damage can be inflicted to the soft tissues of the eyes in a short time. Therefore, if somebody misuses a pen laser, the laboratory supervisor or attendant should be immediately notified.

4.2.5.4 Fire Hazards

In the event of a fire, remain calm and shout in a clear, steady, loud voice to warn your colleagues of the danger and to clear the laboratory. If the fire is small and contained in a beaker or small vessel, you can quickly kill it by smothering the flames with sand, a fire blanket, or fire extinguisher. Organic solvents have the potential to be fuels for fires, so you need to be cautious when handling them and keep them away from ignition sources and open flames. This is also the case when you are handling flammable liquids or gases within the vicinity of an open flame or ignition source. Remember, water should never be used to extinguish a solvent fire. It will only serve to spread the fire, not extinguish it. In the case of solvent fires, a dry chemical type or CO_2 gas fire extinguisher should be used. If the fire event is not small, then the fire alarm should be used and the fire brigade notified immediately.

If your clothes catch on fire, you should shout for help, then lie down on the floor and roll on the affected area to smother and extinguish the fire.

The last thing you should do is run around and feed the fire with air. If a water shower station is nearby, immediately proceed to it and dowse yourself with water. If a fellow student's or colleague's clothes are on fire, immediately wrap the person in a fire blanket and make the person lie on the floor and roll on the affected area to restrict the amount of air feeding the fire. It may be necessary to force the person down on the ground and roll them over to smother the fire. Remember, the time it takes to smother the fire is crucial; prompt action can prevent injuries and may even save lives. Importantly, if you have long hair, make sure it is always tied back to avoid coming in contact with open flames; remember that you can burn just like many other materials. Once the fire is out, immediately notify the laboratory supervisor or attendant and seek medical attention for the injured person.

4.2.6 Hazard Labeling: National Fire Protection Association

In the event of an incident, there will be a response from emergency services, which are expected to identify the hazards and effectively deal with them to resolve the incident. To assist emergency services in identifying the hazards, the National Fire Protection Association (NFPA) developed a standardized system that uses a fire diamond or hazmat diamond signage to provide information about the potential hazard in a concise and informative manner to the fire brigade, emergency services and other response personnel. These hazard labels can be seen in the laboratory; for example, reagent bottles are labeled with hazmat diamond signage, which provides basic and valuable information about the contents in the bottle. The NFPA 704 fire diamond and hazmat diamond are described in the NFPA704 Standard, which is maintained by the NFPA in the Unites States of America. In the laboratory, the most convenient source of NFPA 704 material data and hazard ratings can be found in the Material Safety Data Sheet (MSDS), which is supplied by the manufacturer of the material. The diamond-shaped signage contains four smaller diamonds within it, with each smaller diamond having a particular color designation that represents a specific hazard category.

There are four colors used to indicate the four hazard categories: health (blue), fire/flammability (red), instability/reactivity (yellow), and the fourth diamond is white and is reserved for special hazards, such as water reactivity or the presence of an oxidizer. The health, fire, and instability ratings of a material are also based on their physical and chemical characteristics by a number system. The numbers used in the hazard-rating system consist of 4, 3, 2, 1, or 0, with 4 indicating a severe hazard or an extreme danger and a 0 rating indicating no hazard warning. Examples of each rating are presented in Figures 4.2a and 4.2b.

(a)

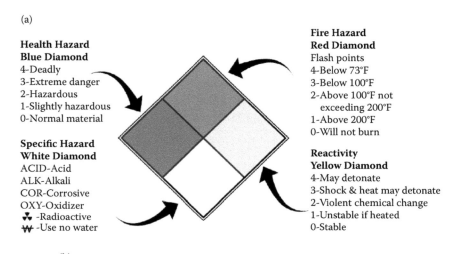

Health Hazard
Blue Diamond
4-Deadly
3-Extreme danger
2-Hazardous
1-Slightly hazardous
0-Normal material

Specific Hazard
White Diamond
ACID-Acid
ALK-Alkali
COR-Corrosive
OXY-Oxidizer
☢ -Radioactive
W -Use no water

Fire Hazard
Red Diamond
Flash points
4-Below 73°F
3-Below 100°F
2-Above 100°F not
  exceeding 200°F
1-Above 200°F
0-Will not burn

Reactivity
Yellow Diamond
4-May detonate
3-Shock & heat may detonate
2-Violent chemical change
1-Unstable if heated
0-Stable

(b)

FIGURE 4.2   (a) NFPA fire and hazard diamond sign. (b) Examples on reagent bottles.

4.2.7 Summary of Important Safety Rules

Laboratory safety is important for both you and your colleagues and can never be taken lightly. It is for both your own protection and the protection of your colleagues, and it is for this reason that you must at all times adhere to the following laboratory rules:

1. YOU MUST

   - Wear safety glasses at all times

   - Tie back long hair

   - Report all accidents immediately

   - Wear closed-in or covered shoes (no thongs or sandals)

   - Wear lab coats at all times

   - Return equipment and materials where they belong after use

2. YOU MUST NOT

   - Eat

   - Smoke

   - Drink

   - Work unsupervised

   - Use volatile solvents near open flames

   - Show any inappropriate behavior in the laboratory—you will be expelled from the laboratory session

3. TIPS AT A GLANCE

   - Think safety first.

   - Read the MSDSs for all chemicals you will be using.

   - Know emergency responses.

   - Know what chemicals you are working with and how to safely handle them.

- When working with chemicals, always start with the smallest possible amount you need.

- Store and handle hazardous materials safely.

- Above all, if you do not know ... ASK.

4.2.8 Safety in Teaching Laboratories

The following safety rules and procedures are specifically designed for teaching laboratories and are intended to protect you and others working with you. Acceptable, safe working habits will greatly complement your academic growth and provide you with the prerequisites required by future employers. Remember, from an employer's point of view, an employee who works safely is a productive employee. Because safety is an important aspect of laboratory work, any individual who breaches the safety rules that follow, thereby endangering themselves or other students, will be excluded from the laboratory and will not be readmitted until they are prepared to follow the safety rules.

In general, you are required to wear safety glasses and wear a laboratory coat (or an approved equivalent) and suitable footwear. Sandals, thongs, and "open" designer shoes are not considered acceptable footwear. The following is a list of safety guidelines designed to protect you and give you a safe experience while learning experimental skills in a teaching laboratory [2]:

1. When in a school laboratory, you must wear appropriate safety glasses.

2. Toxic, fuming, vaporizing, or aggressive chemicals shall only be handled in fume cupboards or similar enclosed apparatus.

3. If you are unsure about any substance or techniques you will be working with, ASK your supervisor for advice BEFORE you proceed. It is always safer to regard all substances as potentially hazardous and apply all appropriate safe working practices.

4. No food or drink is to be brought into the laboratory and there will be no drinking from laboratory taps or glassware. If you require any refreshments, you must leave the laboratory after removing your lab coat, washing your hands, and advising your supervisor of your intentions.

5. Smoking is strictly not permitted within any laboratory, academic building, or building precincts where smoke may enter buildings via doorways, windows, or ventilation grills.

6. Before commencing any laboratory procedure and prior to leaving the laboratory, you should wash your hands with soap and water to remove all contaminants.

7. You are personally responsible for any personal articles you bring with you into the laboratory. Other than necessary writing materials, calculator, and the like that are required during the laboratory session, personal articles should not be brought into the laboratory. You will need to store bags and other personal articles in relevant storage areas.

8. Avoid any hand-to-mouth movements; this will reduce the risk of contamination or poisoning.

9. Long hair must be tied up or confined when undertaking laboratory work.

10. Be aware of fire exits and other escape routes that you may need to use in the event of an evacuation. Also, be aware of fire extinguishers and fire blanket locations.

11. If you have a spillage of any chemical, alert your demonstrator immediately so that appropriate action can be taken.

12. If you suffer a splash from any chemical solution, immediately wash the affected parts with water and notify the supervisor because further treatment may be required.

13. While in a laboratory, you are required to act responsibly at all times. "Practical jokes" and general misbehavior will not be tolerated in a laboratory, and you will be excluded from the laboratory.

14. Many of the instruments and glassware that you will be using are sensitive, delicate, and expensive. If you are unsure how to use them, ASK your laboratory supervisor or attendant. If you discover any defective equipment, you should immediately report it to your supervisor or laboratory attendant.

15. You are required to dispose of your own experimental materials as appropriate. Your supervisor will advise you about the various

disposal procedures relevant to the materials you have been using. Remember to tidy up and clean your bench space before you depart the laboratory.

16. Importantly, report all accidents, incidents, or "near misses" to the laboratory supervisor or attendant.

4.2.9 Evacuation Procedures

In the event of an emergency, you may be required to evacuate the laboratory. In this event, the following procedures should be followed: When the **alarm signal** sounds, you are expected to evacuate the building using the closest safe exit route and proceed to a designated assembly location (or muster point) in an orderly manner. Do not panic and inform your fellow students about the alarm signal before you evacuate. If there is no immediate danger, a floor warden or your supervisor will direct you to do the following:

1. Turn off all electrical and gas equipment.
2. Secure all chemicals.
3. Secure all personal belongings.
4. Evacuate to assembly location (muster point).

During the evacuation, follow the instructions of the fire warden or your supervisor. You will be directed to use staircases; do not use the elevator. Walk in an orderly way and do not run; do not take refuge in the toilets or go up to the roof of the building. When you have reached the assembly location, a "roll call" will be conducted, so do not wander off or it will be assumed that you are still in the building.

In the event that you encounter serious difficulties during the evacuation, stay in the laboratory; the floor warden will be the last to leave the area after checking that all the rooms and laboratories have been cleared and will be able to assist you. After the evacuation, remain at the assembly location and do not reenter the building until advised to do so by the floor warden or your laboratory supervisor.

4.3 SCIENTIFIC REPORT WRITING

4.3.1 Introduction

Experimentation is of fundamental importance and is a crucial part for making discoveries in science and engineering. However, just as important is the need to accurately record the results of the experimental work and present them in an informative and effective format. Experimental studies have been the cornerstone of science since the early 1600s and are an integral part of the scientific method used in science and engineering today. Whatever scientific techniques are used to investigate a particular phenomenon, proper reporting procedures must be followed so that all the data are captured and recorded. The findings from an experimental investigation and the procedures used to obtain the results are usually published in scientific journals or industrial reports. Therefore, before a report or an article can be written, appropriate and accurate records of the experimental work must be recorded during the research. Thus, keeping a log of the experimental work by means of a laboratory notebook can provide a proper structured and easily readable account of the experiment and the procedures used. These notes can then be translated into a properly formatted, structured, and well-written report or article, which can be used to disseminate the results of the study.

The layout of a report has a strong impact on its readability and acceptance by the reader. In addition, a scientific report is written in the third person, so instead of stating, "I measured the mass of lead block to be 15 grams," the accepted way to write it is "the mass of lead block was measured and found to be 15 grams." Thus, the mass of the lead block is presented without actually stating who was the actual person measuring the mass of the block. In scientific reporting, the person who carries out the experiment is not the focus of the report; in any case, the authors are listed below the title on the cover page, and the acknowledgments can be used to mention particular people or organizations that have assisted in the study.

There are several ways to write a laboratory report, and its format can depend on the field of research you are working in. This is because each research field has developed a particular format to suit its type of research and reporting procedures. To reduce the ambiguity and confusion between various report formats, a generalized format is presented in Figure 4.3. The report is usually divided into six sections: (1) experimental aims (or hypothesis being tested); (2) study background; (3) experimental procedure (with a materials section); (4) results obtained; (5) a discussion of the results; and finally (6) a brief conclusion.

4.3.2 Getting Started

Before you attend the laboratory to do an experimental exercise or project, it is a good idea to fully read *all* the background material and procedures needed to undertake the exercise or project in the laboratory session.

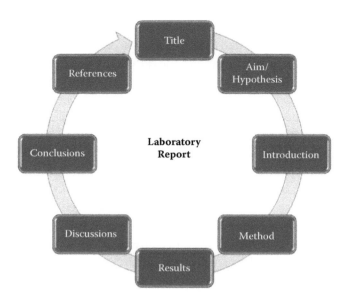

FIGURE 4.3   General components of a scientific laboratory report.

By doing this, you will have an overview of the experimental work that has to be done, and you can plan ahead. Generally, during the experiment you will be engaged in collecting data, and you should be prepared to collect all of the bits of information (and any modifications or unforeseen incidents). Therefore, you will need a proper laboratory handbook to write down your observations and perform any necessary calculations during the experiment. It is common to see science students record their findings on loose sheets of paper; they believe that it is more convenient, but more often than not, they generally misplace the sheets and lose the data. Remember to write down all the data in your notebook neatly and in a format you can easily understand because you will need to translate your notes into a report.

When it comes to writing your report, you should write grammatically correct sentences. Try to be concise and do not confuse the reader by using long sentences. Generally, the passive tense is recommended in science reports, and personal pronouns such as I, you, and we are not to be used. Be objective in your writing and do not use words such as *wonderful* or *great* in your report. Use the spelling and grammar checker function in your word processor to assist you in writing a legible report without grammatical errors. If you have a smartphone with a built-in camera, you can easily capture images during different stages of the experiment that can be used later to assist you in writing the report.

The results of scientific investigations must be made available for other scientists and engineers to reproduce (confirm the results) and for use for further research developments. This process is achieved by publishing; journal articles are a medium through which scientists have the opportunity to communicate their research outcomes. It is not surprising to find that a general format has evolved over the years that is similar to the laboratory report. It is this style that you are recommended to use when writing your laboratory reports and projects. The format consists of the following headings: Title, Abstract, Introduction, Materials and Methods, Results, Discussion, Conclusion, Acknowledgments, and References. Each of these headings (except for Title) is discussed in more detail in the following sections.

4.3.3 Report Format

4.3.3.1 Abstract

The abstract is a brief summary of the research carried out, its significance, and the results of the study. It provides an overview of the research so that the reader can see that the report matches his or her research interests. It should be brief and is usually between 150 and 250 words long depending on the journal. Although the abstract appears just under the article title, it is usually the last section that is written. This is because it must be a self-contained short description of the whole paper and briefly state the results.

4.3.3.2 Introduction

The introduction to the laboratory report is designed to give the reader a summary of the experimental work undertaken and the research outcomes. Because it is important to grab the attention of the reader from the very first, it needs a good starting paragraph. The next paragraph usually describes any background research that is relevant to the experimental work that was carried out as part of this report. To write a successful introduction, you must strike a balance between providing sufficient background information needed by the reader, where your work fits, and the explanation of the experimental work carried out. If you get the balance wrong, the chances are you will lose your reader.

If you have to refer to someone else's work, give a brief description of the important points that are relevant to your report; the reader does not want a detailed discussion of the referenced work. Importantly, a good introduction contains an outline of the current understanding or knowledge in the field that pertains to the results of your work. A general rule for a good introduction is that it cannot be much greater than 20% of the complete report.

4.3.3.3 Materials and Methods

This section is important and provides details of all the materials used, such as purity, batch, and other manufacturer's details. Remember only SI units can be used in writing reports for an international audience. The section also details all experimental procedures used and laboratory equipment used to characterize the synthesized materials. The section must be laid out in logical order because other scientists need to be able to replicate the procedure using the same materials and laboratory equipment, thus verifying the accuracy of your experimental results. If you are following a standard technique, you need to reference the technique so that the reader has the opportunity to access details of the technique and the procedures followed in performing the experimental work. In many cases, a standard laboratory technique may be modified to suit a particular application; in this case, any modifications must be clearly defined.

4.3.3.4 Results

Once you have completed the background (Introduction and Materials and Methods), you are in the position to report the results of your experimental work. It is not necessary to flood the reader with all the details of the experiment or all the data you have obtained. Instead, you can present sufficient representative data. It is also a good idea, if possible, to present your data concisely in tables because it is a neat way to present various sets of results at the same time.

Also, presenting your data in a graphical form is an excellent method to capture the whole story; it allows you to display large quantities of data at once. Like the old adage says, "A picture is worth a thousand words," this is also true of a graph. Generally, we are all attracted to pictures over words, so a graphical presentation is an effective method of conveying large amounts of data rapidly. Other advantages of using graphics are the ability to determine quantities such as the gradient of the line of best fit and the maximum of a peak value.

Importantly, the impact of experimental uncertainties in the raw data that can contribute to small differences in the calculated values can be shown in a proper graph. Remember that quantifying the level of uncertainty will help in discussing your observed results in the Discussion section.

4.3.3.5 Discussion

In the Discussion section, you will interpret the results derived from the experimental studies. Many aspects can be discussed, but you need to focus on the major issues, not the trifles. In some cases, the results will not reflect the general idea or particular hypothesis on which the experiment was based. In these cases, you should identify why the results are different and identify any shortcoming with the experimental procedure that may have influenced the results. After having performed the experiment, the results may even provide you with a better way of achieving your experimental aims; this can also be discussed in this section. In some report formats, both the Results and Discussion sections are combined. The format you adopt is usually decided by where the report or article is presented; companies and publishers usually follow their own format, and it is wise to check with them on what they require.

4.3.3.6 Conclusion

The conclusion generally focuses on the original aims of the experiment and summarizes the various parts of the report into coherent and cohesive discussion. It should include any implications resulting from the research outcomes and make recommendations for any future work. In many cases, you will be involved in determining a well-known physical or chemical parameter, such as the value of g ms^{-2} (acceleration caused by the earth's gravitational pull). In this case, the value is well known and can be referenced to a data book or textbook. In this case, you would compare the experimental value and the accepted value and comment on the validity of your experiment. The conclusion should be the logical ending of the report; it cannot bring in new data or information because that would distract from the work done in the report. The conclusion must be concise and as a general rule is no greater than 20% of the report length. It should also finish with a final statement composed of a few well-chosen sentences that close the report with a positive impact on the reader.

4.3.3.7 References

Bearing in mind that university learning is very much based on the previous research of others, these sources of knowledge must be acknowledged and proper information provided so that other researchers can be directed to these original sources. To do this, you must cite your sources of information. Failure to do this is a serious offense. Plagiarism is defined as the act of using someone else's ideas and information and making it your own.

This is dishonest and is not acceptable behavior. This behavior is a serious issue and breach of scientific etiquette. If you commit plagiarism in your report writing, you will fail your experimental session, and this may ultimately result in failing your course. So, please be careful about citing all your references properly.

Generally, references can be useful in providing background information or an overview of the field, so a recent book or review of the topic can be used to provide information on the subject area. There are different ways to acknowledge or cite your sources of information. The most commonly used method is the Harvard system, which is a popular author-date-based format, but there are many other styles that can be used. It is recommended that you check which system is preferred by your course coordinator, and if more information is needed, consult your university librarian. Remember, you must be consistent in the referencing format you choose for your laboratory report, making sure that the references are properly referenced to the appropriate text in each section of your report. One of the easiest ways to do this is to add a superscript next to the point in the text that is related to the reference. For example, this book uses square brackets containing a number to indicate the numbered reference item. A typical reference listing should include the name of the each author, the full title of the article or chapter, year of publication, the journal or book title, the volume number (in bold or underlined), and the sequence of page numbers. It should be pointed out that publishers have their own formatting styles for referencing, so you need to consult them or else the article or report will be rejected. For example, a famous STM (scanning tunneling microscopy) paper that described how the atomic resolving power of the technique was used to confirm theoretical surface science models. When this paper was cited using the American Chemical Society style (the format used in this book), it appeared as follows: Binnig, G.; Rohrer, H.; Gerber, Ch.; Weibel, E. 7 x 7 reconstruction on Si(111) resolved in real space. *Phys. Rev. Lett.* **1983**, *50*, 120–123.

4.3.4 Closing Remarks

Scientific report writing is an important and essential skill that must be mastered by all professionals working in science and engineering fields. It is important to use proper English spelling and grammar correctly to make the report legible and easy to read, follow, and understand. Remember, using incorrect spelling and bad grammar will undermine the creditability of your report or article. It must also be written in the third person.

The presentation and style of the report or article are important factors because the first impression will have a significant impact on the reader. Allow enough space between each section, be consistent in your formatting, and always proofread your material before submission. Careless mistakes in the text and formatting will undermine your creditability as a competent researcher.

Furthermore, if other people, companies, and institutions have had some input into the report they need to be acknowledged as well. The best way to inform the reader about the contributions of others is in the Acknowledgment section, which comes before the References section. For example, if you have been sponsored under a scholarship or if a laboratory assistant has helped you in some part of the experimental work, then it is appropriate to offer thanks for these contributions. As a final note, if there is more experimental data that need to be part of the report but would crowd it, then these data can be placed in appendices (with headings such as Appendix A, Appendix B, etc.), at the end of the report. For further information regarding scientific writing and how to present scientific data, the student will find the work of both Kirkup [3] and Mathews et al. [4] helpful texts.

REFERENCES

1. Harwood, L. M.; Moody, C. J. *Experimental Organic Chemistry: Principles and Practice*; Blackwell Scientific: Oxford, UK, 1989.
2. Malati, M. A. *Experimental Inorganic/Physical Chemistry: An Investigative, Integrated Approach to Practical Project Work*; Horwood: Chichester, UK, 1999.
3. Kirkup, L. *Experimental Methods: An Introduction to the Analysis and Presentation of Data*; Wiley-VCH: Weinheim, Germany, January **1996**, 1–9.
4. Matthews, J. R.; Matthews, R. W. *Successful Scientific Writing*; Cambridge University Press: Melbourne, 2008.

CHAPTER 5

Nanotechnology Laboratories

5.1 SYNTHESIS OF GOLD NANOPARTICLES BY A WET CHEMICAL METHOD

5.1.1 Aim

The aim is to synthesize gold nanoparticles (NPs) by a fast, wet chemical reduction method.

5.1.2 Introduction

Gold (Au) was a highly sought-after precious metal long before the beginning of recorded history. It has influenced human history in various ways (e.g., art, monetary systems) and has even affected human migration (i.e., driven many gold rushes and settlements across several continents). It is still highly used in jewelry and fashion. Gold is a dense, soft, shiny, malleable, and extremely ductile metal. Pure gold has a bright yellow luster and color, which it maintains without oxidizing in air or water. Gold has an atomic number of 79, and chemically the element is considered a transition metal. The electron configuration of gold is $[Xe]4f^{14}5d^{10}6s^1$. Because gold has been with us for such a long time, it is no surprise that we have tried and tested this material for many uses and even its biocompatibility with human tissues and possible uses in medicine. The early Egyptians and several other civilizations used gold, and there is ample evidence to show its use as dental implants to fix tooth decay. Even today, there is an entire field of medicine devoted to the application of gold compounds—called chrysotherapy—for the treatment of ailments such as arthritis and other illnesses.

Gold nanoparticles (Au NPs) are of great interest today for various nanoelectronic applications, such as bio-/chemosensors, nanoelectronic components, biological tags, and even catalysts. The renowned British experimentalist Michael Faraday made some of the first Au NPs in 1857 while studying the interaction of light with metallic particles "very minute in their dimensions." About 100 years later, Turkevich used electron microscopic investigations to reveal that the ruby-colored colloids of gold made in a similar way as Faraday's preparative routes produced particles of gold with average sizes in the range of 6 ± 2 nm. Since then, there has been constant exploration into nanogold research and development (R&D). Today, many rapid diagnostic tests rely solely on the ability of bioconjugated Au NPs to detect specialized molecules for the early detection of diseases. Other research efforts have shown that Au NPs can also help to identify cancer growths and other diseases when used as biomarkers.

There are several ways to synthesize Au NPs, and the easiest route is through the aqueous method because there are many gold water-soluble salts. Chloroauric acid ($HAuCl_4$), a yellow/orange solid, is such a compound; it has been used in many preparative routes to manufacture Au NPs. After dissolution in water, the $[AuCl_4]^-$ ions are reacted with a reducing agent. This causes the reducing agent to react with the Au^{3+} ions of the $AuCl_4^-$ to give a neutral Au^0 atom. As the concentration of these neutral atoms increases, the solution becomes supersaturated, and the Au^0 atoms aggregate to form Au NPs. At this stage, particles of various sizes can be created, but ultimately a generalized shape and size is obtained depending on the reaction conditions. As NPs, the surface energy is high, and the Au NPs tend to agglomerate further. To stop this process, a stabilizing agent or capping agent is added that will stick to the surface of the metallic NP. Several compounds exist that can act as capping agents, and these can be easily fine-tuned to assist in the bioconjugation of the Au NPS in further applications. The optical property of gold NPs is tunable through the visible and near-infrared region of the spectrum as a function of NP size, shape, aggregation state, and local environment. The size and shape of the gold NPs and the viewing conditions (transmitted light or reflected light) determine the color of the gold solution that is ultimately obtained.

In this laboratory session, chloroauric acid is used as a source of gold ions, and for the reducing agent, borohydride ions are used. One interesting property of colloidal particles, because of their shape and size, is that they scatter white light in a process called the Tyndall effect.

Named after the nineteenth-century physicist John Tyndall, the effect is the process of light being scattering and reflected by colloidal particles or NPs in suspension. The presence of a colloidal suspension can be easily detected by the scattering/reflection of a laser beam from the NPs as the beam of light passes through the solution. In contrast, when the laser beam is shined through a normal solution (i.e., gold chloride solution) without colloids or NPs, the beam passes through without scattering. The Tyndall effect can only be used to determine if a colloid/NP suspension is present, but it cannot determine the concentration of the NPs in that solution. Thus, it acts as a qualitative tool in the rapid determination of Au NPs in this instance because the human eye cannot directly see individual NPs in the solution.

Tyndall Effect

When the laser beam is shone through a normal solution (i.e., gold chloride solution) without colloids or nanoparticles, the beam passes through without being scattered (A)

The presence of a colloidal suspension can be easily detected by the scattering/reflection of a laser beam from the nanoparticles as beam of light passes through the solution (B)

Tyndall effect showing the scattering of laser light because of the presence of gold nanoparticles.

5.1.3 Key Concepts

1. Sodium borohydride is a strong reducing agent and can reduce gold(III) ions to Au^0 and in turn Au NPs. Other reducing agents include sodium hypophosphite, hydrazine, hydrazine sulfate, and aminoboranes.

2. Gold NPs, having a large surface area and associated high surface energy, tend to agglomerate to achieve stability. To prevent this agglomeration, sodium citrate is used as a capping agent to keep Au NPs separated in solution.

3. The size and shape of the Au NPs and the viewing conditions (transmitted light or scattered light) determine the color of the gold solution we see.

4. The presence of a colloidal suspension can be detected by the scattering of a laser beam from the NPs/nanostructures (Tyndall effect).

5.1.4 Experimental

5.1.4.1 Materials/Reagents

Source of gold ions	0.01 M chloroauric acid	$(HAuCl_4)$
Capping agent	0.01 M sodium citrate	$(C_6H_5Na_3O_7)$
Reducing agent	0.10 M sodium borohydride	$(NaBH_4)$
Milli-Q® water		(H_2O)

5.1.4.2 Glassware/Equipment
Clean glassware is essential for this experiment.

Glass vials (20 mL)	Laser pointer
Measuring cylinder (10 mL)	Pasteur pipettes
Beakers (50 mL)	Kimwipes®
Micropipettes (200–1000 μL)	Digital camera
Micropipette tips (200–1000 μL)	

5.1.5 Special Safety Precautions

1. Before starting the experiment, please look up the Materials Safety Data Sheets for all the chemicals that will be used. All chemicals used should be regarded as irritants and toxic; avoid inhalation of any chemicals.

2. Nitrile gloves, laboratory coats, and safety eyewear should be worn at all times while in the laboratory.

3. Clean glassware is essential for this experiment as minute contaminants can affect the end results.

4. The laser beam can be harmful to the eye, so avoid looking straight into it. Never shine a laser beam at another person because it can result in eye damage.

5.1.6 Procedure: Preparation of Gold Nanoparticles by a Wet Chemical Method

- For each experimental procedure, note any observable changes in the solutions, such as color change, precipitate formation, effervescence.

- For each observation, explain what is occurring during the reaction process.

- Take digital photographs of the experimental reactions that can be included in your final report.

Part I: Concentrated Au NPs with capping agent (sodium citrate)

Part II: Concentrated Au NPs without capping agent (sodium citrate)

Part III: Dilute Au NPs with capping agent (sodium citrate)

Part IV: Dilute Au NPs without capping agent (sodium citrate)

5.1.6.1 Part I: Concentrated Gold Nanoparticles with Sodium Citrate

Part I: Concentrated Gold Nanoparticles with Sodium Citrate

Procedure	Observations	Inferences/ Comments
To 10 mL of Milli-Q water add • 1.0 mL of 0.01 M HAuCl$_4$ • 1.0 mL of 0.01 M sodium citrate Stir/mix well.		
Use a laser pointer to show whether there is scattering or nonscattering of the laser beam in the Au solution.		
Carefully add 1.0 mL of 0.1 M NaBH$_4$. *Note*: Be careful as a vigorous reaction may occur.		
Use a laser pointer again to confirm if there is formation of Au NPs. If NPs are present, there should be a scattering effect when the laser beam is shined through the solution, making them visible.		

Note any observable changes in the gold NP reaction, such as color change or precipitate formation.

5.1.6.2 Part II: Concentrated Gold Nanoparticles without Sodium Citrate

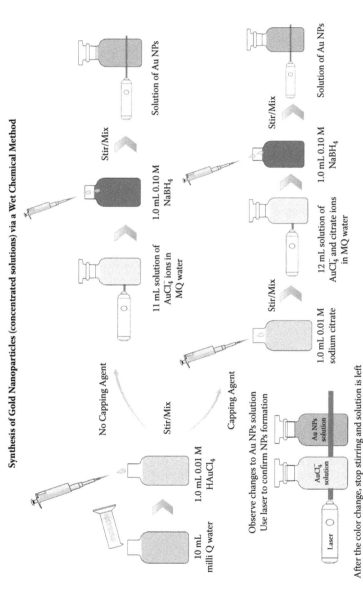

Synthesis of gold nanoparticles with concentrated solution of gold chloride with and without capping agent.

Part II: Concentrated Gold Nanoparticles without Sodium Citrate

Procedure	Observations	Inferences/ Comments
To 10 mL of Milli-Q water, add 1.0 mL of 0.01 M HAuCl$_4$ and mix well.		
Use a laser pointer to show whether there is scattering or nonscattering of the laser beam in the Au solution.		
Carefully add 1.0 mL of 0.1 M NaBH$_4$. *Note:* Be careful as a vigorous reaction may occur.		
Use a laser pointer again to confirm if there is formation of Au NPs. If NPs are present, there should be a scattering effect when the laser beam is shined through the solution, making them visible.		

Note any observable changes in the gold NP reaction, such as color change or precipitate formation.

5.1.6.3 Part III: Dilute Gold Nanoparticles with Sodium Citrate

Part III: Dilute Gold Nanoparticles with Sodium Citrate

Procedure	Observations	Inferences/ Comments
To 10 mL of Milli-Q water add • 0.5 mL of 0.01 M HAuCl$_4$ • 1.0 mL of 0.01 M sodium citrate Stir/mix well.		
Use a laser pointer to show whether there is scattering or nonscattering of the laser beam in the Au solution.		
Carefully add 1.0 mL of 0.1 M NaBH$_4$. *Note:* Be careful as a vigorous reaction may occur.		
Use a laser pointer again to confirm if there is formation of Au NPs. If NPs are present, there should be a scattering effect when the laser beam is shined through the solution, making them visible.		

Note any observable changes in the gold NP reaction such as color change, precipitate formation

5.1.6.4 Part IV: Dilute Gold Nanoparticles without Sodium Citrate

Synthesis of Gold Nanoparticles (diluted solutions) via a Wet Chemical Method

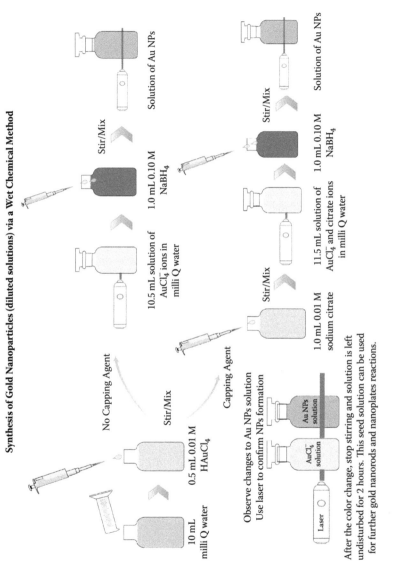

Synthesis of gold nanoparticles with a diluted solution of gold chloride with and without capping agent.

Part IV: Dilute Gold Nanoparticles without Sodium Citrate

Procedure	Observations	Inferences/Comments
To 10 mL of Milli-Q water, add 0.5 mL of 0.01 M HAuCl$_4$ and mix well.		
Use a laser pointer to show whether there is scattering or nonscattering of the laser beam in the Au solution.		
Carefully add 1.0 mL of 0.1 M NaBH$_4$. *Note:* Be careful as a vigorous reaction may occur.		
Use a laser pointer again to confirm if there is formation of Au NPs. If NPs are present, there should be a scattering effect when the laser beam is shined through the solution, making them visible.		

Note any observable changes in the gold NP reaction, such as color change or precipitate formation.

5.1.7 Characterization of Gold Nanoparticles

Once the Au NPs are synthesized, these samples can be kept for further analysis at a further stage with nanocharacterization tools or for further experimentation in the nanoscience project section. The morphologies, shape, and sizes of the Au NPs can be analyzed by these techniques:

- field emission scanning electron microscopy (FE-SEM)
- transmission microscopy (TEM)
- atomic force microscopy (AFM)
- scanning tunneling microscopy (STM)

The color of Au NPs can be analyzed by a normal UV-Vis spectrophotometer.

FURTHER READING MATERIAL

1. Turkevich, J. Colloidal gold. Part II. *Gold Bull.* **1985,** *18* (4), 125–131.
2. Hayat, M. A. *Colloidal Gold: Principles, Methods, and Applications*; Elsevier: New York, 2012.
3. Sharma, V.; Park, K.; Srinivasarao, M. Colloidal dispersion of gold nanorods: Historical background, optical properties, seed-mediated synthesis, shape separation and self-assembly. *Mater. Sci. Eng. R* **2009,** *65* (1), 1–38.
4. Reddy, V. R. Gold nanoparticles: synthesis and applications. *Synlett* **2006,** (11), 1791–1792.

5.2 BIOSYNTHESIS OF ECO-FRIENDLY SILVER NANOPARTICLES

5.2.1 Aim

The aim is to synthesize NPs of silver by a green, surfactant-free, eco-friendly method.

5.2.2 Introduction

Silver (Ag) is a soft, translucent gray transition metal with an atomic number of 47. This element has the highest thermal and electrical conductivity of all the metallic elements found so far on Earth. The electron configuration of silver is $[Kr]4d^{10}5s^1$. It can be found in pure form or alloyed with other elements. It is considered a precious metal and is used for coins and bullions as part of monetary systems; it can be found in jewelry and ornaments. The photographic film industry has relied on the photochemical reactivity of silver and silver halide nanoclusters for many years to generate photographs.

Because of its close association with human history in several applications, silver has also been tested and used in dentistry in the form of amalgams with mercury for fillings to repair tooth decay. In addition, silver compounds have long been known for their antimicrobial activity. With the advent of a plentiful supply of antibiotics after World War II, there has been a major shift away from metal ions and their antimicrobial effects (oligodynamic effects) until recently. As microbes can evolve, resistant bacterial strains, for example, are capable of handling the latest most potent generations of antibiotics, and coupled with their ability to pass on this resistance to others, it makes them formidable pathogens. Thus, there has been resurgence in the manufacture of nanosilver not only as an antibacterial but also for potential application against fungi, viruses, and other pathogenic species. There is currently a drive to make nanosilver and apply it as coating for various medical components to stop infections.

Green chemistry is one of the new branches of chemistry, and it involves the design of products and processes that reduce or eliminate the use or generation of hazardous substances. Green synthetic routes for manufacturing NPs and nanostructures are an emerging branch of nanotechnology as the biomolecules around us are safer generally and offer a cost-effective alternative in many cases. For example, today one would be rather reluctant to undertake Michael Faraday's 1857 method of reducing gold chloride with red phosphorus in a volatile, toxic carbon disulfide solution as a technique to create Au NPs. In many conventional methods, there is a tendency to use expensive chemicals and processes that use toxic materials that present hazards such as environmental toxicity and carcinogenic activity. There has been a push toward an alternative pathway of minimizing the use and production of hazardous materials in chemical research.

The term *green chemistry* (or sustainable chemistry) was coined by Paul Anastas in 1991 while at the Environmental Protection Agency (EPA); with John Warner, the 12 principles of green chemistry were developed. One facet of this approach has been to increasingly use plants, fungi, and bacteria to generate beneficial compounds. Given that these living organisms are in our environment already, it provides a wide and efficient platform to create molecules and compounds efficiently. For instance, millions of diabetics worldwide need daily injections of insulin, a hormone that regulates sugar levels in their blood. Unfortunately, in this case, the body does not produce enough of the hormone, and it has to be injected daily and monitored. The supply of insulin has always been an issue as the demand outgrows the supply from animal sources. In 1982, an altered form of insulin derived from *Escherichia coli* bacteria was approved for diabetics, and this type of insulin partly alleviated the supply issue. Thus, by employing bacteria in this instance and through genetic engineering we were able to help manage diabetes.

Sustainable or green technique pathways that create materials utilizing relatively nontoxic chemicals to create nanomaterial are well favored and are welcomed avenues of R&D efforts around the world. Following initial reports showing the feasibility of reducing silver ions to Ag NPs, there has been a general move to explore plant extract as a means of reducing silver to produce NPs and nanostructures of this metal. In some plants, the acidic components can easily aid the reduction of the metallic ions.

Furthermore, these studies showed that Ag NPs created this way possess good antimicrobial activity. The fact that no capping agent or templating agent is needed makes this chemical route an attractive one. Today, there have been many reports about Ag NPs made by plant extracts and those plants with known medicinal properties. For instance, the biogenesis of Ag NPs by extracts such as those from the neem (*Azadirachta indica*), geranium leaves (*Pelargonium graveolens*), and alfalfa (*Medicago sativa*) has already been proven, and the list of plants capable of this reducing effect on silver ions is increasing. Even benign solutions such as tea and coffee extracts have the ability to create NPs of silver and palladium at room temperature. Nadagouda et al. showed that 20- to 60-nm particles were easily obtained by this simple technique. Similar to the previous experiments, you will synthesize Ag NPs using a plant/leaf extract in Part I and one from green tea leaves in Part II.

5.2.3 Key Concepts

1. Green chemistry principles of using nontoxic raw materials for synthesis

2. Surfactant-free Ag NPs

3. Stable Ag NPs by a reduction method

5.2.4 Experimental
5.2.4.1 Materials/Reagents

Source of silver ions	0.01 M silver nitrate	$(AgNO_3)$
Reducing agent	Plant leaf extract	
	Green tea extract	
	Milli-Q water	

Types of plant/leaf extracts that can be used:

- Rose leaf petals
- Taro leaves
- Geranium leaves
- Alfalfa

5.2.4.2 Glassware/Equipment

Measuring cylinder (50 mL)	Micropipette (200–1000 µL)
Glass vials (20–50 mL)	Micropipette tips (200–1000 µL)
Beakers (100 mL)	Buchner funnel
Schott bottle (50 mL)	Mortar and pestle
Conical flask (100 mL)	Spatula
Magnetic stirrer	Hot plate
Filter paper	Laser pointer
Kimwipes	Digital camera
Analytical balance	

5.2.5 Safety Precautions

1. Before starting the experiment, look up Materials Safety Data Sheets for all the chemicals that will be used.

2. All chemicals used should be regarded as irritant and toxic; avoid inhalation.

3. Wear gloves, laboratory coats, and safety eyewear at all times while in the laboratory.

4. Silver nitrate can stain, so make sure you wear gloves.

5. **Clean glassware is essential for this experiment**.

6. The laser beam pointer can be harmful to the eyes, so avoid looking straight into the laser beam. Never shine a laser beam at another person.

5.2.6 Procedure

- For each experimental procedure note, any observable changes in the Ag NP reaction, such as color change, precipitate formation.

- For each observation, explain what is occurring during the reaction process.

- Take digital photographs of the experimental reactions to be included in your final report.

Part I: Biosynthesis of Ag NPs using plant/leaf extracts

Part II: Biosynthesis of Ag NPs using green tea

5.2.6.1 Part I: Biosynthesis of Silver Nanoparticles Using Plant/Leaf Extracts

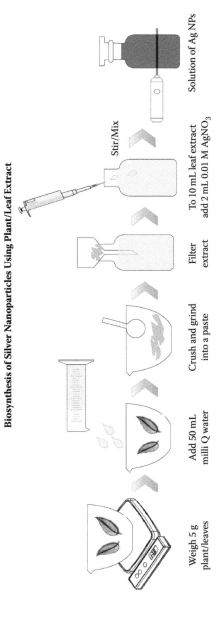

Biosynthesis of Silver Nanoparticles Using Plant/Leaf Extract

Weigh 5 g plant/leaves	Add 50 mL milli Q water	Crush and grind into a paste	Filter extract	To 10 mL leaf extract add 2 mL 0.01 M AgNO$_3$	Stir/Mix

Solution of Ag NPs

Observe the Ag NPs Solution Every 15 Minutes for 60 Minutes

Observe changes to Ag NPs solution
Use laser to confirm NPs formation

Biosynthesis of silver nanoparticles using plant/leaf extracts.

Part I: Biosynthesis of Silver Nanoparticles Using Plant/Leaf Extract

Procedure	Observations	Inference/ Comments
To prepare the plant/leaf extract, weigh 5 g of plant/leaves and add 50 mL of Milli-Q water; grind into a paste using a mortar and pestle. Filter the plant/leaf extract into a Schott bottle.		
To 10.0 mL of plant/leaf extract add 2 mL of 0.01 M $AgNO_3$ and mix/stir.		
Use a laser pointer to show whether there is scattering or nonscattering of the laser beam in the Ag solution.		
Observe the Ag NP solution after **15 minutes.**		
Observe the Ag NP solution after **30 minutes.**		
Observe the Ag NP solution after **45 minutes.**		
Observe the Ag NP solution after **60 minutes.**		
Use a laser pointer again to confirm if there is formation of Ag NPs.		

Note any observable changes in the silver NP reaction, such as color change or precipitate formation.

5.2.6.2 Part II: Biosynthesis of Silver Nanoparticles Using Green Tea

Biosynthesis of Silver Nanoparticles Using Green Tea Extract

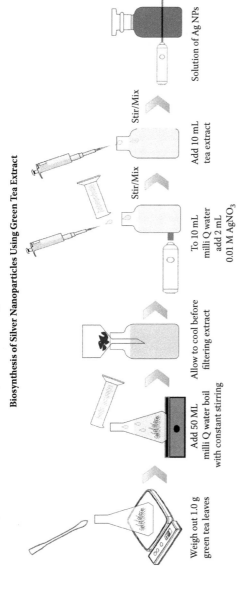

Weigh out 1.0 g green tea leaves → Add 50 ML milli Q water boil with constant stirring → Allow to cool before filtering extract → To 10 mL milli Q water add 2 mL 0.01 M AgNO₃ — Stir/Mix → Add 10 mL tea extract — Stir/Mix → Solution of Ag NPs

Observe the Ag NPs Solution Every 15 Minutes for 60 Minutes

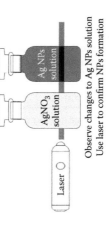

Laser | AgNO₃ solution | Ag NPs solution

Observe changes to Ag NPs solution
Use laser to confirm NPs formation

Biosynthesis of silver nanoparticles using a green tea extract.

Part II: Biosynthesis of Silver Nanoparticles Using Green Tea Leaf Extract

Procedure	Observations	Inference/ Comments
To prepare the green tea extract: Weigh 1.0 g of green tea leaves and boil using 50.0 mL of Milli-Q water with constant stirring.		
Once tea extract comes to a boil, allow to cool and filter into a Schott bottle.		
To 10.0 mL of Milli-Q water add 2 mL of 0.01 M AgNO$_3$ and stir.		
Use a laser pointer to show whether there is scattering or nonscattering of the laser beam in the Ag solution.		
Add 10.0 mL of tea extract and mix/stir.		
Observe the Ag NP solution after **15 minutes.**		
Observe the Ag NP solution after **30 minutes.**		
Observe the Ag NP solution after **45 minutes.**		
Observe the Ag NP solution after **60 minutes.**		
Use a laser pointer again to confirm if there is formation of Ag NPs.		

Note any observable changes in the silver NP reaction, such as color change or precipitate formation.

5.2.7 Characterization of Silver Nanoparticles

Once the Ag NPs have been synthesized, these samples can be kept for further analysis.

The morphology and shape of the NPs can be analyzed by these techniques:

- FE-SEM
- TEM
- AFM
- STM

The color and size of the Ag NPs can be analyzed by a UV-Vis spectrophotometer.

FURTHER READING MATERIAL

1. Safaepour, M.; Shahverdi, A. R.; Shahverdi, H. R.; Khorramizadeh, M. R.; Gohari, A. R. Green synthesis of small silver nanoparticles using Geraniol and its cytotoxicity against Fibrosarcoma-Wehi 164. *Avicenna. J. Med. Biotechnol.* **2009**, *1* (2), 111.

2. Nadagouda, M. N.; Varma, R. S. Green synthesis of silver and palladium nanoparticles at room temperature using coffee and tea extract. *Green Chem.* **2008**, *10* (8), 859–862.

3. Sun, Q.; Cai, X.; Li, J.; Zheng, M.; Chen, Z.; Yu, C.-P. Green synthesis of silver nanoparticles using tea leaf extract and evaluation of their stability and antibacterial activity. *Colloids Surf. A Physicochem. Eng. Aspects* **2014**, *444*, 226–231.

4. Bar, H.; Bhui, D. K.; Sahoo, G. P.; Sarkar, P.; Pyne, S.; Misra, A. Green synthesis of silver nanoparticles using seed extract of *Jatropha curcas*. *Colloids Surf. A Physicochem. Eng. Aspects* **2009**, *348* (1), 212–216.

5.3 SYNTHESIS OF ZINC SULFIDE NANOPARTICLES BY A REVERSE MICELLE METHOD

5.3.1 Aim

The aim is to use micro-/nanocavities of a reverse micelle method to prepare semiconductor zinc sulfide NPs.

5.3.2 Introduction

Colloidal semiconductor NPs have been studied for many years because of their novel optical properties. These materials exhibit a particular phenomenon called quantum confinement when their sizes are comparable to the diameter of bulk excitons (bound state of an electron and hole). In this case, the continuum of states is broken into discrete states in these materials. In particular, tiny particles called quantum dots (QDs) have attracted considerable interest from many areas. These QD are mainly from the chalcogenide family: sulfide and selenide of metallic elements like zinc or cadmium. In addition, InP and PbS QDs have been synthesized by other research teams. The size ranges are about 2–10 nm in diameter, and they emit light when irradiated or a current is passed into them. QDs produce light in a similar way that atoms and electrons are excited to a higher level and then as they drop they emit light of a certain wavelength and depending on the band gap. It was discovered that smaller QDs have a larger gap between energy levels, so more energy is needed to promote the excited electrons. In turn, the emitted light will reflect this energy content, so the wavelength is shorter than one emitted from a larger QD. In the biological field, these QDs, with their tunable optical properties, can be engineered as specific fluorophores (they emit a certain color) to tag certain cells or organelles within a cell. This makes them remarkable bioprobes as these nanocrystals have long-term photostability, and their use not only allows live cell imaging but also provides dynamic studies of cellular material.

The particle size relationship to the band gap is a direct reflection of quantum confinement in semiconductors: As the particle size decreases, the band gap increases. The difference in the properties of NPs is generally caused by two factors: the increase in the surface-to-volume ratio and the change in the electronic structure of the material caused by quantum confinement effects. As the particle size decreases, the quantum confinement effects begin to have greater influence on the particle characteristics. Wide-band-gap II–VI semiconductor materials such as sulfides have gained much attention because of their wide applications in the fields of light-emitting devices, solar cells, sensors, and optical recording materials.

Zinc sulfide (ZnS) is a wide-band-gap semiconductor, and it has two structural forms. The cubic form has a band gap of 3.54 eV at room temperature, whereas the hexagonal form has a band gap of 3.91 eV. This material is commercially used in a variety of applications, including optical coatings, solid-state solar cell windows, thin-film electroluminescent devices, photoconductors, field effect transistors, sensors, light-emitting applications, and even toys that light up in the dark.

Synthesis of regular and uniform NPs is the foundation of the field of nanotechnology and nanoscience. To achieve such uniformity, robust manufacturing steps are required to ensure the correct morphology and reproducibility. There are many methodologies available for synthesizing ZnS powders, such as laser ablation, electrochemical fabrication, solvothermal methods, reverse micelle methods, and chemical vapor deposition (CVD). One such technique is the reverse micelle method, which you will try in this laboratory session. The rationale behind this technique is simple: To synthesize ultrafine NPs (a small quantity of matter), one need only react small quantities of the reactants to produce such small particles. In this respect, the aqueous reactants are dispersed and are enclosed in micellar structures called aqueous (water-filled) nano-/micropools or nano-/microreactors (refer to the first figure for this section). These reactors, containing very small quantities, move in a continuous phase of oil or any organic medium, with their motion governed by Brownian motion. When these nanoreactors collide, they exchange their contents, and the zinc ions can react directly with the sulfide ions to form ZnS NPs (refer to the second and third figures this section). As more collisions occur, the NPs grow to an optimum size. One of the advantages of this technique is the uniformity of the resulting nanostructures, and this technique does not require extreme conditions or high temperatures. So, the concept of encapsulating a small concentration of reactant A (Zn^{2+}) and reactant B (S^{2-}) (refer to the fourth figure) in an oil continuous medium and making them react is an attractive option. This is the basis of the reverse micelle method of producing NPs. The advantage of using this methodology is primarily that it ensures a low polydiversity of fine particles. It is a low-energy input method because it works at room temperature or close to room temperature, and color variation is a strong indication of a favorable reaction. This method is widely used to make semiconductor QDs, such as those of ZnS, CdS, or CdTe.

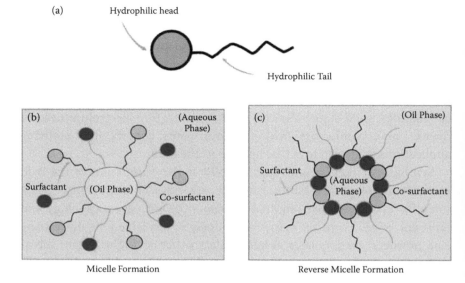

Schematic of a surfactant molecule (a) and representation of a micelle (b) and reverse micelle (c) with the main components.

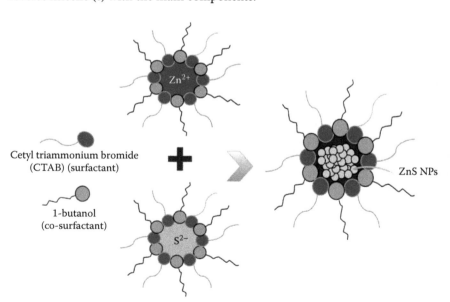

Schematic of a reverse micellar reaction to produce ZnS NPs.

5.3.3 Key Concepts

1. Use of small-volume reverse micelles to manufacture NPs

2. Surfactants, water, and a continuous oil phase create stable reverse micelles

3. Micro-/nanoreactors engineer stable NPs and nanostructures

5.3.4 Experimental

5.3.4.1 Materials/Reagents

Source of zinc ions	0.05 M	Zinc sulfate	$(ZnSO_4)$
Source of sulfide ions	0.05 M	Sodium sulfide	(Na_2S)
Surfactant	CTAB	Cetyl triammonium bromide,	$(C_{19}H_{42}BrN)$
Cosurfactant		1-Butanol	(C_4H_9OH)
Oil phase		n-Hexane	(C_6H_{14})
Glass drying agent		Acetone	(C_3H_6O)
		Milli-Q water	(H_2O)

5.3.4.2 Glassware/Equipment
Clean dry glassware is essential for this experiment.

Glass vials (20–50 mL)	Spatula
Beakers (100 mL)	Analytical balance
Micropipettes (200–1000 µL)	Ultraviolet light
Micropipette tips (200–1000 µL)	Digital camera
Pasteur pipettes	Kimwipes
Laser pointer	

5.3.5 Special Safety Precautions

1. Before starting the experiment, look up the Materials Safety Data Sheets for all the chemicals that will be used. All chemicals used should be regarded as irritating and toxic; avoid inhalation.

2. Wear gloves, laboratory coats, and safety eyewear at all times in the laboratory.

3. Clean and DRY glassware is essential for this experiment.

4. Because of the chemical vapors of the oil phase, work in a fume hood.

5.3.6 Procedure: Synthesis of Zinc Sulfide Nanocrystals via Reverse Micelle Method

- For each experimental procedure, note any observable changes in the reaction, such as a color change or precipitate formation.

- For each observation, explain what was occurring during the reaction process.

- Take digital photographs of the experimental reactions to be included in your final report.

Part I: Preparation of the primary oil phase slurry

Part II: Preparation of the solution fractions

Part III: Preparation of concentrated ZnS nanocrystals

Part IV: Preparation of diluted ZnS nanocrystals

Zinc sulfide nanocrystals can be manufactured via the reaction of zinc ions (Zn^{2+}) and sulfide ions (S^{2-}) in aqueous solutions.

The reverse micelle solution **(A)** is made up of an aqueous solution of

- 0.05 M zinc sulfate ($ZnSO_4$) **(source of Zn^{2+})**

- CTAB (surfactant)

- 1-Butanol (cosurfactant)

- A continuous oil phase of hexane

The reverse micelle **(B)** solution is made up of an aqueous solution of

- 0.05 M sodium sulfide (Na_2S) **(source of S^{2-})**

- CTAB (surfactant)

- 1-Butanol (cosurfactant)

- A continuous oil phase of hexane

5.3.6.1 Part I: Preparation of the Primary Oil Phase Slurry

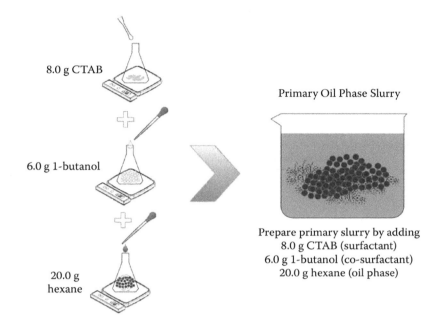

8.0 g CTAB

6.0 g 1-butanol

20.0 g
hexane

Primary Oil Phase Slurry

Prepare primary slurry by adding
8.0 g CTAB (surfactant)
6.0 g 1-butanol (co-surfactant)
20.0 g hexane (oil phase)

Preparation of primary slurry prior to addition of metallic and sulfide ions.

This mixture must be **kept dry at ALL TIMES**; it must not contain any water at this stage.

- Clean DRY glassware is essential for this part of the experiment.

- Acetone can be used to promote the removal of water from your glassware.

Part I: Preparation of Primary Oil Phase Slurry System

Procedure	Observations	Inference/Comments
Prepare the primary oil phase slurry system by adding 8.0 g of surfactant CTAB, 6 g of cosurfactant 1-butanol, and 20.0 g of oil phase of hexane in a glass beaker.		

Keep this mixture; *Note and observe the mixture at this stage.*

5.3.6.2 Part II: Preparation of the Solution Fractions

Part II: Preparation of the Solution Fractions

Separate the primary oil slurry into exactly 4 fractions in 4 labeled glass vials.	
Labeled Vial	**Vial Will Contain**
A1	Fraction will contain 800 μL Zn^{2+} solution.
A2	Fraction will contain 200 μL Zn^{2+} solution.
B1	Fraction will contain 800 μL S^{2-} solution.
B2	Fraction will contain 200 μL S^{2-} solution.

5.3.6.3 Part III: Preparation of Concentrated ZnS Nanocrystals

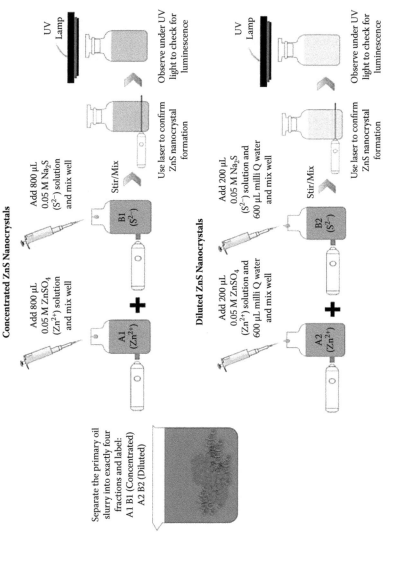

Preparation of ZnS nanocrystals by a reverse micelle method.

Part III: Preparation of Concentrated ZnS Nanocrystals

Procedure	Observations	Inference/Comments
To the A1 fraction, add 800 μL of 0.05 M $ZnSO_4$ (Zn^{2+}) solution.		
To the B1 fraction, add 800 μL of 0.05 M Na_2S (S^{2-}) solution.		
Use a laser pointer to show whether there is scattering or nonscattering of the laser beam in each of the fractions.		
Now you have prepared two similar systems with nanopools of exactly the same size moving about because of Brownian motion. *Note any changes that occur when Zn^{2+} and S^{2-} are added to the A1 and B1 fractions.*		
Mix the A1 fraction and B1 fractions together.		
Note any observable changes when the 2 fractions (Zn^{2+} and S^{2-}) are added together. How long does it take to react?		
Use a laser pointer to confirm the formation of ZnS nanocrystals.		
Observe under the UV light to check for any luminescence.		

5.3.6.4 Part IV: Preparation of Diluted ZnS Nanocrystals

Part IV: Preparation of Diluted ZnS Nanocrystals

Procedure	Observations	Inference/Comments
To the A2 fraction, add 200 μL of 0.05 M $ZnSO_4$ (Zn^{2+}) solution and 600 μL of Milli-Q water.		
To the B2 fraction, add 200 μL of 0.05 M Na_2S (S^{2-}) solution 600 μL of Milli-Q water.		
Use a laser pointer to show whether there is scattering or nonscattering of the laser beam in each of the fractions.		
Now you have prepared two similar systems with nanopools of exactly the same size moving about because of Brownian motion. *Note any changes that occur when Zn^{2+} and S^{2-} are added to the A2 and B2 fractions.*		
Mix the A2 fraction and B2 fractions together.		
Note any observable changes when the 2 fractions (Zn^{2+} and S^{2-}) are added together. How long does it take to react?		

Continued

Procedure	Observations	Inference/Comments
Use a laser pointer to confirm the formation of ZnS nanocrystals.		
Observe under the UV light to check for any luminescence.		

5.3.7 Characterization of Zinc Sulfide Nanocrystals

Once the zinc sulfide nanocrystals have been synthesized, these samples can be kept for further analysis.

The morphology and shape of the nanocrystals can be analyzed by these techniques:

- FE-SEM
- TEM
- AFM

The color of the ZnS nanocrystals can be observed under a UV light.

The color and size of the ZnS nanocrystals can be analyzed by a UV-Vis spectrophotometer.

FURTHER READING MATERIAL

1. Agostiano, A.; Catalano, M.; Curri, M.; Della Monica, M.; Manna, L.; Vasanelli, L. Synthesis and structural characterisation of CdS nanoparticles prepared in a four-components "water-in-oil" microemulsion. *Micron* **2000**, *31* (3), 253–258.
2. Loukanov, A. R.; Dushkin, C. D.; Papazova, K. I.; Kirov, A. V.; Abrashev, M. V.; Adachi, E. Photoluminescence depending on the ZnS shell thickness of CdS/ZnS core-shell semiconductor nanoparticles. *Colloids Surf. A Physicochem. Eng. Aspects* **2004**, *245* (1), 9–14.
3. Tang, H.; Xu, G.; Weng, L.; Pan, L.; Wang, L. Luminescence and photophysical properties of colloidal ZnS nanoparticles. *Acta Mater.* **2004**, *52* (6), 1489–1494.
4. Sfihi, H.; Takahashi, H.; Sato, W.; Isobe, T. Photoluminescence and chemical properties of ZnS: Mn^{2+} nanocrystal powder synthesized in the AOT reverse micelles modified with lauryl phosphate. *J. Alloys Compd.* **2006**, *424* (1), 187–192.

5.4 SYNTHESIS OF FLUORESCENT CARBON NANOPARTICLES FROM CANDLE SOOT

5.4.1 Aim

The aim of this experiment is to produce fluorescent carbon nanoparticles (CNPs) from candle soot.

5.4.2 Introduction

Ever since the discovery of buckminsterfullerenes or buckyballs in 1985, many other types of carbon nanostructures, such as carbon nanotubes (CNTs), nanofilaments, nanocapsules, and recently CNPs from the C_n family, have been synthesized and studied by a number of research groups worldwide. Carbon spheres can be divided into three categories: (1) the C_n family and well-graphitized onion-like structures (diameters between 2 and 20 nm); (2) less-graphitized nanosize spheres (diameters between 50 nm and 1 μm); and (3) carbon beads (diameters between 1 and several micrometers). Among the various allotropic forms of carbon, CNPs have become increasingly important and have been gaining steady interest because of their novel properties. Professor Andre Geim and Professor Konstantin Novoselov were awarded the Nobel Prize in Physics for their groundbreaking experiments with graphene in 2010.

Carbon-based nanomaterials, which include carbon nanotubes, fullerenes, and nanofibers, have promising applications in nanotechnology, biosensing, and drug delivery. Recently, CNPs—a new class of carbon-based nanomaterials with interesting photoluminescence properties—were prepared and isolated. Fluorescent NPs have attracted increasing attention because of their promising applications, from electro-optics to bionanotechnology. To date, typical photoluminescent particles have been developed from compounds of lead, cadmium, and silicon. But, these materials also have raised concerns over potential toxicity, environmental harm, and poor photostability. Compared to traditional QDs and organic dyes, photoluminescent carbon nanomaterials (CNPs) are superior in chemical inertness and lower toxicity. The emergence of photoluminescent carbon-based nanomaterials has presented exciting opportunities for searching for fluorescent nanomaterials.

The emergence of fluorescent CNPs shows high potential in biological labeling, bioimaging, and other different optoelectronic device applications. These CNPs are generally biocompatible and chemically inert, which presents advantages over conventional cadmium-based QDs. The range of experimental techniques to synthesize carbon nanospheres/CNPs include arc discharge;

pyrolysis of hydrocarbons (styrene, toluene, benzene, hexane, and ethene); CVD; laser ablation of graphite; thermal decomposition of organic compounds; electro-oxidation of graphite; and oxidation of candle soot. Among all these synthetic methods, the soot-based approach is simple and straightforward. However, the quantum yield of fluorescent CNPs can be low (<0.1%).

In this laboratory, you will generate carbon nano- and microparticles by the simple combustion of a candle and differentiate between the several fractions using a simple chromatographic technique. Thin-layer chromatography (TLC) is used often in organic chemistry or the pharmaceutical industry to monitor the progress of a reaction as it separates the various components of a mixture. At the analytical level, it can be used to identify compounds if a standard is known. TLC is made by first spotting a thin layer of stationary phase (silica gel) on an inert backing with the mixture and then placing the spotted plate into a chamber with a volatile solvent (mobile phase). As in most chromatography, each compound has a certain affinity with the respective mobile phase and stationary phase. This is reflected in the speed at which this compound moves under the capillary action of the mobile phase as it moves up the TLC strip. The aim is to separate the fraction into distinct spots that can be identified easily. TLC is widely used because of its low cost, simplicity, speed, and high sensitivity.

Liu and coworkers from Purdue University in Indiana first demonstrated the UV response of oxidized CNPs formed in candle soot. In their study, they found that the fluorescent properties of the candle soot were dependent on the size of the CNPs formed. The candle soot was simply collected by placing a glass plate on top of candles. The soot contained mainly elemental carbon (elemental analysis: C 91.69%, H 1.75%, N 0.12%, and O 4.36%) and was hydrophobic and insoluble in common solvents. However, this soot could be dissolved in alcoholic solutions.

5.4.3 Key Concepts

1. Combustion process to create carbon micro- and nanomaterials

2. Fluorescence properties of carbon materials

3. Separation by a chromatographic technique: TLC

5.4.4 Experimental

5.4.4.1 Materials/Reagents

Ethanol	(CH_3CH_2OH)
Hexane	(C_6H_{14})
Milli-Q water	

5.4.4.2 Glassware/Equipment

Measuring cylinder (10 mL)	0.2-μm syringe filter
Glass vials (20 mL)	Candle
Beaker (50 mL)	Matches
Microscope slide	Tweezers
Watch glass	Scalpel
Pasteur pipette	Ultrasound bath
Silica TLC plate (2.5 × 7.5 cm)	UV lamp (325 nm)
Capillary tubes	Laser pointer
Melting point tubes	Digital camera
Syringe (5 mL)	Kimwipes

5.4.5 Special Safety Precautions

- Before starting the experiment, look up the Materials Safety Data Sheets for all chemicals that will be used. All chemicals used should be regarded as irritating and toxic; avoid inhalation.

- Wear gloves, laboratory coats, and safety eyewear at all times while in the laboratory and take full precautions during the course of experiment.

- Care must be taken during the collection of soot from the burning candle.

- Care must be taken not to burn yourself with the open flame and hot wax.

5.4.6 Procedure

- For each experimental procedure, note any observable changes, such as a color change before and after sonication of the soot and before and after filtering of the carbon soot and the color of the soot under the UV light of the filtered and unfiltered samples.

- Take digital photographs of the experimental reactions to be included in your final report.

Part I: Synthesis of fluorescent CNPs from candle soot

Part II: Separation of fluorescent CNPs using the TLC method

5.4.6.1 Part I: Synthesis of Fluorescent Carbon Nanoparticles from Candle Soot

Synthesis of Fluorescent Carbon Nanoparticles from Candle Soot

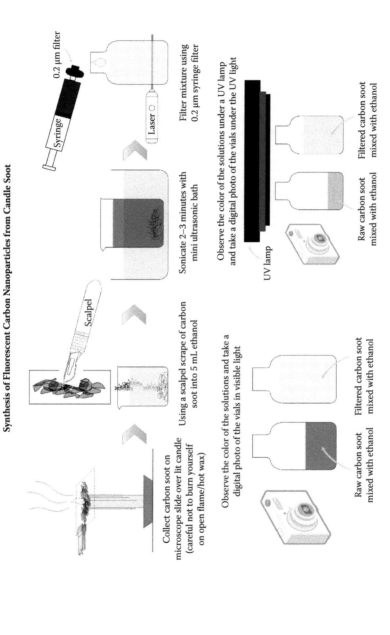

Collect carbon soot on microscope slide over lit candle (careful not to burn yourself on open flame/hot wax)

Using a scalpel scrape of carbon soot into 5 mL ethanol

Sonicate 2–3 minutes with mini ultrasonic bath

Filter mixture using 0.2 μm syringe filter

Observe the color of the solutions and take a digital photo of the vials in visible light

Observe the color of the solutions under a UV lamp and take a digital photo of the vials under the UV light

Raw carbon soot mixed with ethanol

Filtered carbon soot mixed with ethanol

Raw carbon soot mixed with ethanol

Filtered carbon soot mixed with ethanol

Direct synthesis of carbon soot raw material, containing micro- and nanocarbons, from the combustion of a candle.

Part I: Synthesis of Fluorescent Carbon Nanoparticles from Candle Soot

Procedure	Observations	Inference/Comments
Carefully collect the candle soot by placing a microscope slide on top of a candle. Collect the candle soot on two microscope slides. *Care must be taken during the collection of soot from the burning candle.*		
Dissolve the soot in 5 ml of ethanol.		
Filter the mixture into a separate vial using a 0.2-μm syringe filter (1 ml of the raw mixture should to be kept for further analysis). Observe the color of the solution.		
Take a digital picture of the raw carbon soot vial and the filtered carbon soot vial under visible light.		
Use a laser pointer to confirm the presence of CNPs in the filtered vial. If CNPs are present, there should be a scattering effect with the laser beam.		
Place both vials under a UV lamp and observe the color change. Take a digital picture of the vials under the UV light.		

5.4.6.2 Part II: Separation of Fluorescent Carbon Nanoparticles Using the Thin-Layer Chromatographic Method

Thin Layer Chromatography Plate Preparation

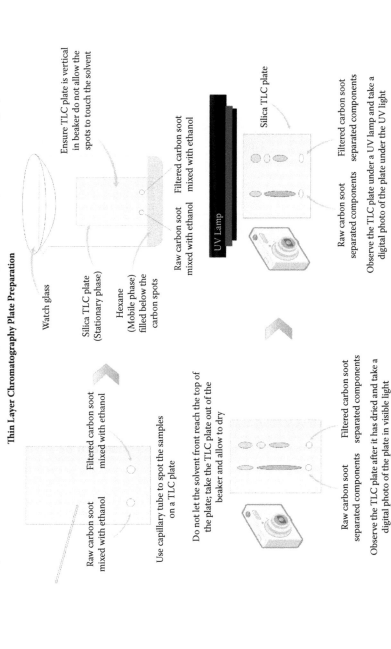

Separation of fluorescent micro-/nanocarbon fractions by thin-layer chromatography.

Part II: Separation of Fluorescent Carbon Nanoparticles Using a Thin-Layer Chromatography Method

Procedure	Observations	Inference/Comments
Take a silica TLC plate (2.5 cm × 7.5 cm) and place a small drop of raw soot mixture and filtered mixture using a capillary tube.		
Place ~ 5 mL of hexane in a beaker; check that the amount added is less than the spots on the TLC plate. Place the TLC plate into the beaker, place a watch glass on the beaker, and allow the solvent to move up the plate (~5 minutes).		If the hexane is more than the spots when the TLC plate is added, the spots will spoil, and the TLC will not work. Also, ensure the TLC plate is placed vertically in the beaker; otherwise, the solvent will not move up properly.
Take the TLC plate out of the beaker before the solvent front reaches the top of the TLC plate and let it dry for 1 minute. Take a digital picture of the TLC plate in the visible light.		Do not let the solvent front flow to the top of the TLC plate; otherwise, the separated components will spoil.
Place the TLC plate under a UV lamp (325 nm) and observe the changes. Take a digital picture of the TLC plate under the UV light.		

5.4.7 Characterization of Carbon Nanoparticles

The CNPs formed can then be characterized further by the following:

- The morphology and shape of the nanocrystals can be analyzed by these techniques:
  - FE-SEM
  - TEM
- The size of the CNPs can be analyzed by a UV-Vis spectrophotometer.

FURTHER READING MATERIAL

1. Liu, H.; Ye, T.; Mao, C. Fluorescent carbon nanoparticles derived from candle soot. *Angew. Chem. Int. Ed.* **2007**, *46* (34), 6473–6475.
2. Sun, Y.-P.; Zhou, B.; Lin, Y.; Wang, W.; Fernando, K. S.; Pathak, P.; Meziani, M. J.; Harruff, B. A.; Wang, X.; Wang, H. Quantum-sized carbon dots for bright and colorful photoluminescence. *J. Am. Chem. Soc.* **2006**, *128* (24), 7756–7757.

3. Jin, Y. Z.; Gao, C.; Hsu, W. K.; Zhu, Y.; Huczko, A.; Bystrzejewski, M.; Roe, M.; Lee, C. Y.; Acquah, S.; Kroto, H. Large-scale synthesis and characterization of carbon spheres prepared by direct pyrolysis of hydrocarbons. *Carbon* **2005**, *43* (9), 1944–1953.

4. Qian, H.-s.; Han, F.-m.; Zhang, B.; Guo, Y.-c.; Yue, J.; Peng, B.-x. Non-catalytic CVD preparation of carbon spheres with a specific size. *Carbon* **2004**, *42* (4), 761–766.

5. Poinern, G. E. J.; Brundavanam, S.; Shah, M.; Laava, I.; Fawcett, D. Photothermal response of CVD synthesized carbon (nano) spheres/aqueous nanofluids for potential application in direct solar absorption collectors: a preliminary investigation. *Nanotechnol. Sci. Appl.* **2012**, *5*, 49–59.

5.5 SYNTHESIS OF ZINC OXIDE NANORODS BY A MICROWAVE METHOD

5.5.1 Aim

The aim is to synthesize ZnO nanorods via a microwave method.

5.5.2 Introduction

Zince oxide (ZnO) is an interesting metallic oxide material with many properties that lead to its use in several applications. Zinc has an $[Ar]3d^{10}4s^2$ electronic configuration, whereas oxygen has a $[He]2s^22p^4$ configuration. It is used as a UV blocker in sunscreen to protect the skin from the sun's UV radiation and as a pigment in some paints. ZnO will reflect both UVA and UVB radiation. ZnO has a wide band gap of 3.37 eV, high excitonic binding energy, and high breakdown strength, so it can be utilized for electronic and photonic devices and in high-frequency applications. Currently, there is increasing interest in nanoscience and nanotechnology to investigate the possibility of nanostructured ZnO for light-emitting diodes (LEDs), among other applications. This would allow for more cost-effective general lighting as it is a much cheaper and efficient technology to produce than incandescent bulbs, and this type of application would have a better environmental impact. Among the semiconducting oxide nanorods, ZnO, with its large exciton binding energy and wide band gap, is one of the materials with the most potential. Research in a variety of fields has already shown that ZnO nanorods are promising candidates that can be used in UV nanolasers, field effect transistors, solar cells, and nanogenerators.

With the advent of the commercial microwave in the 1960s, there has been a steady increase in its use in the common household. Slowly but surely, this method is increasingly being used in laboratories especially for chemical synthesis. Whenever a reaction needs energy to proceed, the input of energy can come in many forms, and the most traditional is thermal energy. Microwave heating is called volume heating because the whole system is heated at the same time by the tremendously fast switching of each water molecule (or other polar molecules present) at gigahertz levels. This allows for a much faster reaction process, and in some cases, the product yields are much higher than from traditional heating. In the nanorealm, microwave synthesis has great potential to fabricate nanomaterials and self-assembled nanostructures at a fast rate. Size-dependent material properties in the 1- to 100-nm range can be significantly different from the material properties relative to molecular and bulk properties of the same material. This size and shape dependency is both an advantage and a challenge for nanoscience and nanotechnology. To maximize a particular property of the nanomaterial, one has to have a narrow size distribution around the related size (and shape). Rapid nucleation in a fluid heated by volume heating is advantageous because it does not rely on conventional thermal gradients like those generated by heating in a conventional beaker or flask on a hot plate. So, there is a definite role for microwave heating in nanomaterial synthesis. In this experiment, ZnO nanorods are synthesized via the microwave method, thus creating ZnO nanostructures rapidly. Please read the following laboratory exercise first, follow the procedures as detailed, and answer the questions that follow. Record all your observations as you proceed with the experiment.

5.5.3 Key Concepts

1. Fabrication of a thin film of UV-shielding material

2. Synthesis of nanomaterials using a conventional microwave

3. Introduction of a different chemical reaction pathway

4. Synthesis of ZnO nanorods

5.5.4 Experimental

5.5.4.1 Materials/Reagents

Ammonium chloride (NH_4Cl)	50 mL	0.02 M
Zinc sulfate ($ZnSO_4$)	40 mL	0.01 M
Urea (CH_4N_2O)	5 mL	0.01 M
Ammonia (NH_3) solution	5 mL	28% w/w
MilliQ water (H_2O)		
Ethanol (C_2H_6O)		

5.5.4.2 Glassware/Equipment
Clean glassware is essential for this experiment.

Parr microwave acid digestion bombs
Microwave
Microscope glass slide or microscope cover glass slide
Measuring cylinder (10 mL)
Measuring cylinder (50 mL)
Pasteur pipettes

Continued

Micropipettes (200–1000 μL)
Micropipette tips (200–1000 μL)
Beakers (250 mL, 50 mL)
UV light/lamp
Optical microscope
Kimwipes
Ultrasonic bath

5.5.5 Special Safety Precautions

1. Before starting the experiment, look up the Materials Safety Data Sheets for all the chemicals that will be used. All chemicals used should be regarded as irritating and toxic; avoid inhalation.

2. Wear gloves, laboratory coats, and safety eyewear at all times while in the laboratory.

3. Clean glassware is essential for this experiment.

4. Special care needs to be taken when handling the microwave acid digestion bombs after placement in the microwave; excess heat is produced. Let the system cool before opening to avoid splashing of hot solution.

5.5.6 Procedure: Preparation of ZnO Nanorods by a Microwave Method

- For each experimental procedure, record any observation regarding the glass slide before and after it has been microwaved using the Parr acid digestion bombs; note any observable changes in the reaction, such as color change or precipitate formation.

- For each observation, explain what was occurring during the reaction process.

- Take digital photographs of the experimental reactions to be included in your final report.

Part I: Preparation of ZnO solutions

Part II: Synthesis of ZnO nanorods by a microwave method

Part III: Observations of ZnO nanorods by UV and optical microscopy

5.5.6.1 Part I: Preparation of ZnO Solutions

Preparation of ZnO Solutions for Synthesis of ZnO Nanorods by Microwave Method

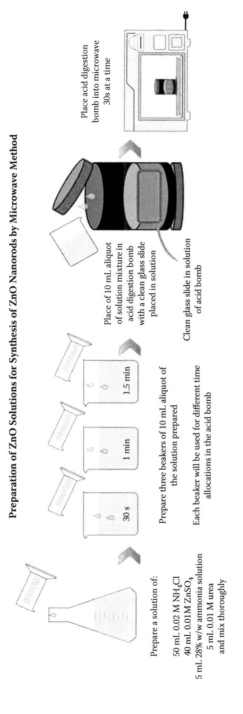

Prepare a solution of:

50 mL 0.02 M NH_4Cl
40 mL 0.01M $ZnSO_4$
5 mL 28% w/w ammonia solution
5 mL 0.01 M urea
and mix thoroughly

30 s   1 min   1.5 min

Prepare three beakers of 10 mL aliquot of the solution prepared

Each beaker will be used for different time allocations in the acid bomb

Place of 10 mL aliquot of solution mixture in acid digestion bomb with a clean glass slide placed in solution

Clean glass slide in solution of acid bomb

Place acid digestion bomb into microwave 30s at a time

Preparation of precursor solution prior to microwave synthesis.

Part I: Preparation of ZnO Solutions

Procedure	Observations	Inference/Comments
Add 50 mL of 0.02 M NH_4Cl, 40 mL of 0.01 M $ZnSO_4$, 5 mL of 0.01 M urea, and 5 mL of 28% w/w ammonia solution to a beaker and mix thoroughly.		
Prepare three beakers of 10-mL aliquots of the prepared solution; each beaker will be used for different time allocations in the microwave digestion bombs.		

Clean the microscope and cover glass slides by placing into a beaker of ethanol and sonicate (5 minutes); then, use Milli-Q water and sonicate (5 minutes) before rinsing with Milli-Q water.

Once glass slides have been cleaned, place into the solution in the cup.

Cover and close the Parr microwave acid digestion bombs and place into the microwave.

5.5.6.2 Part II: Synthesis of ZnO Nanorods by a Microwave Method

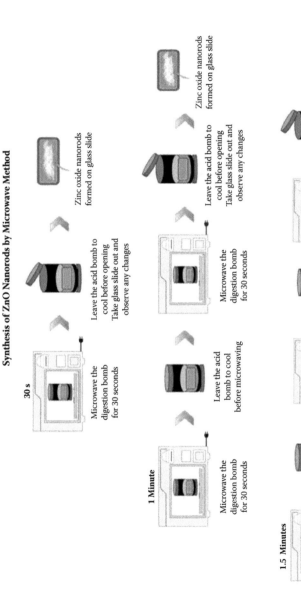

Synthesis of ZnO nanorods by a microwave method.

Part II: Synthesis of ZnO Nanorods by a Microwave Method

Procedure	Observations	Inferences/ Comments
Set the microwave oven to appropriate percentage (100%) for **30 seconds**.		
Note: Please be careful; only use for 30-second intervals or less.		
Once the microwave has finished, leave the digestion bomb to cool before opening. Once the acid digestion bomb solution has cooled, remove the lid and take out the glass slide.		
LET SOLUTIONS IN THE DIGESTION BOMB COOL BEFORE OPENING. Note: Please be careful when opening the bomb because excessive heat that has been produced can cause splashing of solution in the face.		
Repeat using another 10-mL aliquot of the prepared solution.		
This time, place the digestion bomb in the microwave for **1 minute** (2 x 30 seconds).		
Once the bomb has cooled, remove the lid. Observe any changes that have occurred to the glass slide since it was microwaved.		
Repeat using another 10-mL aliquot of the prepared solution.		
This time place the digestion bomb in the microwave for **1.5 minutes** (3 × 30 seconds).		
Once the bomb has cooled, remove the lid. Observe any changes that have occurred to the glass slide since it was microwaved.		

5.5.6.3 Part III: Observations of ZnO Nanorods by Ultraviolet Light and Optical Microscopy

ZnO Nanorods Glass Slide Observed under UV

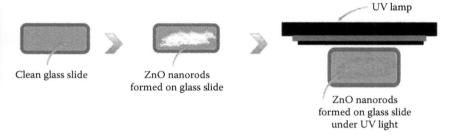

Clean glass slide

ZnO nanorods formed on glass slide

UV lamp

ZnO nanorods formed on glass slide under UV light

ZnO Nanorods Glass Slide Observed under Optical Microscope

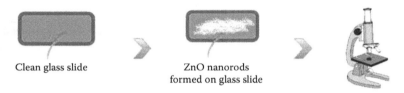

Clean glass slide

ZnO nanorods formed on glass slide

Characterization of thin film of ZnO under UV and by optical microscopy.

Part III: Observations of ZnO Nanorods by UV and Optical Microscopy

Observe any changes that have occurred to the glass slide since it was microwaved. Observe the glass slide under the UV light and optical microscope. Compare the slide to a clean glass slide and record any changes that may have occurred.			
Time Microwaved	**General Observations**	**Observations: UV Light**	**Observations: Optical Microscope**
30 seconds			
60 seconds			
90 seconds			

5.5.7 Characterization of Zinc Oxide Nanorods

Once the zinc oxide nanowires have been synthesized, these samples can be kept for further analysis. The morphology and shape of the nanowires can be analyzed by these techniques:

- FE-SEM

- TEM

- AFM

The color and size of the ZnO nanowires can be analyzed by UV-Vis spectroscopy.

FURTHER READING MATERIAL

1. F.M. Moghaddam, H. Saeidian. Controlled microwave-assisted synthesis of ZnO nanopowder and its catalytic activity for *O*-acylation of alcohol and phenol. *Mater. Sci. Eng., B* **2007**, *139* (2–3), 265–269.
2. N. Takahashi. Simple and rapid synthesis of ZnO by means of a domestic microwave. *Mater. Lett.* **2008**, *62* (10–11), 1652–1654.
3. M. Taylor, B.S. Atri., S. Minhas., P. Bisht. *Developments in Microwave Chemistry*; Evalueserve: Malvern, PA, 2005, 5–25.

5.6 SYNTHESIS OF BIMETALLIC NANOPARTICLES BY WET CHEMICAL METHODS

5.6.1 Aim

The aim is to synthesize bimetallic (Fe@Au and Fe@Ag) NPs by wet chemical methods.

5.6.2 Introduction

Nanometer-size inorganic NPs exhibit size-dependent properties that are different from their bulk material properties; this is mainly because of their large surface-area-to-volume ratio and quantum confinement effects. In the field of nanotechnology and nanoscience, metallic NPs play a vital role, such as in the case of biosensors, immunochemistry, and medical applications. Bimetallic NPs have also been shown to be beneficial in catalysis and related fields. These bimetallic NPs can be made to catalyze the reaction of such environmental pollutants as chlorinated hydrocarbons. Trichloroethene (TCE) used to be a common organic solvent much used in industry, but human exposure to such a chemical has been phased out because of its reported toxic effects on the central nervous system and potential cancer-causing effects. In 2001, Elliot and Zhang from Lehigh University in Pennsylvania showed that iron/palladium 100- to 200-nm NPs could easily break down TCE and other chlorinated aliphatic hydrocarbons. In this field test, a low dosage of 1.7 kg of the iron/palladium NPs was injected into a contaminated well, and over 4 weeks, the efficiency of TCE reduction was up to 96%. In this case, the iron was easily oxidizable as it was in contact with a less-active noble metal; the bimetallic NPs are an effective catalyst for TCE reduction as well as other chlorinated ethanes, benzenes, and PCBs (polychlorinated biphenyls). Given the known toxicity of this type of chemical, it is indeed refreshing to find that a bimetallic nanomaterial can be engineered to fight presently contaminated sites around the world.

Much research still needs to be done in the case of bimetallic NPs as the arrangements of two metals are not guaranteed to be homogeneous; there are cases of preferred distribution of one metal inside the bimetallic system. In some cases, a nanosize core is created, and if it is made of a magnetic material, this bimetallic NP allows a magnetic nanocomposite material with special properties to be engineered. Thus, applications, such as in high-density storage, are possible, as are other biomedical applications, such as protein purification. The physical separation of the NPs with magnetic cores once they are formed is simple and can be achieved easily with a simple laboratory magnet. The structure of bimetallic combinations depends mainly on the preparation conditions and the miscibility of the two components. Therefore, bimetallic combinations exhibit either a core-shell structure or an alloy structure depending on the preparation conditions.

In the present experiment, the aim is to make Fe NPs first by wet chemical methods before coating these *in situ* with gold atoms to form Fe@Au bimetallic NPs. The second part of this experiment involves synthesizing the Fe NPs before coating them with silver atoms to form Fe@Ag NPs. Record all your observations as you proceed with the experiment.

5.6.3 Key Concepts

1. Sodium borohydride is a strong reducing agent and reduces iron ions into Fe NPs quickly. Other popular reducing agents include sodium hypophosphite, hydrazine sulfate, and aminoboranes.

2. Fe NPs, with a large surface area and associated high surface energy, tend to agglomerate to achieve stability. To prevent this agglomeration, sodium citrate is used as a capping agent so that two NPs of iron do not come into direct contact and start to agglomerate.

3. Bimetallic NPs are of great interest as they have novel properties and can lead to different applications. Bimetallic particles with a magnetic core can be used for physical separation of magnetic materials.

5.6.4 Experimental

5.6.4.1 Materials/Reagents

Source of gold ions	0.01 M chloroauric acid ($HAuCl_4$)
Source of iron ions	0.05 M iron(III) chloride ($FeCl_3$)
Source silver ions	0.01 M silver nitrate ($AgNO_3$)
Capping agent for iron NPs	0.50 M sodium citrate ($C_6H_5Na_3O_7$)
Capping agent for gold and silver NPs	0.20 M sodium citrate
Reducing agent	0.05 M sodium borohydride ($NaBH_4$)
Milli-Q water	

5.6.4.2 Glassware/Equipment

Clean glassware is essential for this experiment.

Measuring cylinder (10 mL)
Glass vials (20 mL)
Pasteur pipette

Continued

UV lamp (325 nm)

Magnet

Micropipettes (200–1000 μL)

Micropipette tips (200–1000 μL)

Beakers (50 mL)

Digital camera

Laser pointer

5.6.5 Special Safety Precautions

1. Before starting the experiment, look up the Materials Safety Data Sheets for all the chemicals that will be used. All chemicals used should be regarded as irritating and toxic; avoid inhalation.

2. Wear gloves, laboratory coats, and safety eyewear at all times while in the laboratory.

3. Clean glassware is essential for this experiment.

4. Silver nitrate can stain, so wear gloves

5. The laser beam pointer can be harmful to the eyes, so avoid looking straight into the laser beam. Never shine a laser beam at another person.

5.6.6 Procedure: Synthesis of Bimetallic (Fe@Au, Fe@Ag) Nanoparticles by a Wet Chemical Method

- For each experimental procedure note, any observable changes in the reaction, such as a color change or precipitate formation.

- For each observation, explain what was occurring during the reaction process.

- Take digital photographs of the experimental reactions to be included in your final report.

Part I: Synthesizing Fe@Au bimetallic NPs with capping agent

Part II: Synthesizing Fe@Ag bimetallic NPs with capping agent

5.6.6.1 Part I: Synthesizing Fe@Au Bimetallic Nanoparticles with Capping Agent

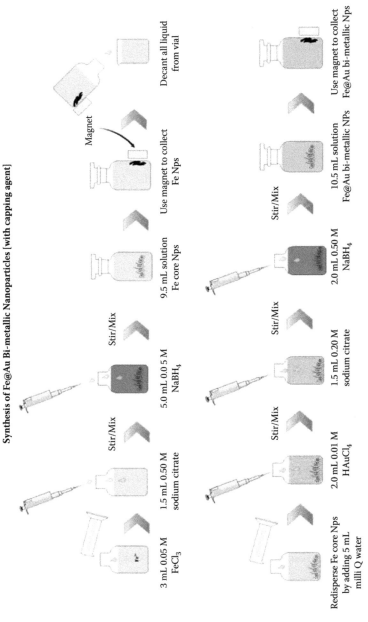

Synthesis of Fe@Au Bi-metallic Nanoparticles [with capping agent]

3 mL 0.05 M FeCl₃ — Stir/Mix — 1.5 mL 0.50 M sodium citrate — Stir/Mix — 5.0 mL 0.05 M NaBH₄ — 9.5 mL solution Fe core Nps — Use magnet to collect Fe Nps — Magnet — Decant all liquid from vial

Redisperse Fe core Nps by adding 5 mL milli Q water — 2.0 mL 0.01 M HAuCl₄ — Stir/Mix — 1.5 mL 0.20 M sodium citrate — Stir/Mix — 2.0 M 0.50 M NaBH₄ — 10.5 mL solution Fe@Au bi-metallic NPs — Use magnet to collect Fe@Au bi-metallic Nps

After color change, stop stirring, magnet is left next to vial to collect Fe@Au bi-metallic Nps

Synthesis of bimetallic Fe@Au NPs by a fast reduction method in the presence of a capping agent.

Part I: Fe@Au Bimetallic Nanoparticles with Capping Agent

Procedure	Observations	Inference/ Comments
To 3 mL of 0.05 M $FeCl_3$, add 1.5 mL of 0.50 M sodium citrate and 5 mL of 0.05 M $NaBH_4$; mix thoroughly to form iron core NPs. *Note:* Be careful as a vigorous reaction may occur.		
Once the reaction is complete, collect the iron core NPs using a laboratory magnet.		
Decant all liquid from the vial and add 5 mL of Milli-Q water to redisperse the iron magnetic cores. To this solution, add 2 mL of 0.01 M $HAuCl_4$, 1.5 mL of 0.20 M sodium citrate, and 2.0 mL of 0.05 M $NaBH_4$; mix thoroughly. *Note:* Be careful as a vigorous reaction may occur.		
Once the solution is mixed thoroughly, leave the vial next to the magnet.		

Note any observable changes in the reaction, such as a color change or precipitate formation.

5.6.6.2 Part II: Synthesizing Fe@Ag Bimetallic Nanoparticles with Capping Agent

Synthesis of Fe@Ag Bi-metallic Nanoparticles [with capping agent]

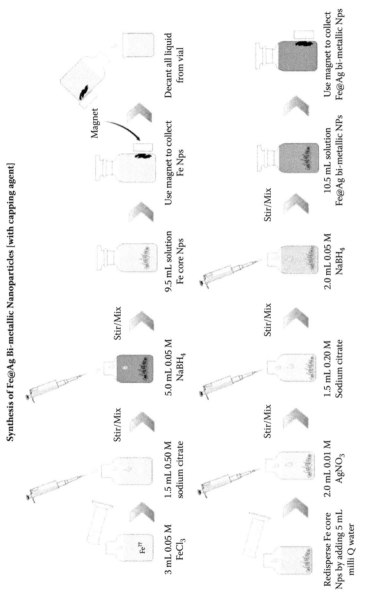

After color change, stop stirring, magnet is left next to vial to collect Fe@Ag bi-metallic Nps

Synthesis of bimetallic Fe@Ag NPs by a fast reduction method in the presence of a capping agent.

Part II: Fe@Ag Bimetallic Nanoparticles with Capping Agent

Procedure	Observations	Inference/ Comments
To 3 mL of 0.05 M FeCl$_3$, add 1.5 mL of 0.50 M sodium citrate and 5 mL of 0.05 M NaBH$_4$; mix thoroughly to form iron core NPs. *Note:* Be careful as a vigorous reaction may occur.		
Once the reaction is complete, collect the iron core NPs using a laboratory magnet.		
Decant all liquid from the vial and add 5 mL of Milli-Q water to redisperse the iron magnetic cores. To this solution, add 2 mL of 0.01 M AgNO$_3$, 1.5 mL of 0.20 M sodium citrate, and 2.0 mL 0.05 M NaBH$_4$; mix thoroughly. *Note:* Be careful as a vigorous reaction may occur.		
Once the solution is mixed thoroughly, leave the vial next to the magnet		

Note any observable changes in the reaction, such as a color change or precipitate formation.

5.6.7 Characterization of Fe@Au and Fe@Ag Bimetallic Nanoparticles

Once the bimetallic NPs have been synthesized, these samples can be kept for further analysis.

The morphology and shape of the NPs can be analyzed by these techniques:

- FE-SEM
- TEM
- AFM

FURTHER READING MATERIAL

1. Elliott, D. W.; Zhang, W.-X. Field assessment of nanoscale bimetallic particles for groundwater treatment. *Environ. Sci. Technol.* **2001,** 35 (24), 4922–4926.
2. Lin, J.; Zhou, W.; Kumbhar, A.; Wiemann, J.; Fang, J.; Carpenter, E.; O'Connor, C. Gold-coated iron (Fe@Au) nanoparticles: synthesis, characterization, and magnetic field-induced self-assembly. *J. Solid State Chem.* **2001,** 159 (1), 26–31.
3. Berger, P.; Adelman, N. B.; Beckman, K. J.; Campbell, D. J.; Ellis, A. B.; Lisensky, G. C. Preparation and properties of an aqueous ferrofluid. *J. Chem. Educ.* **1999,** 76 (7), 943–948.

5.7 SYNTHESIS OF POLYMERIC NANOPARTICLES BY A MODIFIED VERSION OF THE SPONTANEOUS EMULSIFICATION SOLVENT DIFFUSION METHOD

5.7.1 Aim

The aim is to synthesize polymer, poly(lactide-co-glycolide) acid (PLGA), NPs by a modified version of the spontaneous emulsification solvent diffusion (SESD) method.

5.7.2 Introduction

Biodegradable polymeric systems have received considerable attention with their use in applications such as medical devices (sutures), implants, prosthetics, orthopedic repair material, dental devices, pharmaceutical applications, and gene therapy. Increasing attention has been paid to these biodegradable polymeric drug carriers because they can be used to deliver a range of pharmaceutical drug agents for targeting numerous sites of the body. These NPs can be utilized for delivering hydrophilic drugs, hydrophobic drugs, proteins, vaccines, and biological macromolecules. They can be formulated for targeting sites such as the lymphatic system, brain, arterial walls, lungs, liver, and spleen and for even longer-term systematic circulation within the body. Development of ideal drug delivery systems requires the active agent to be encapsulated into a polymer matrix; therefore, an appropriate selection of the polymer matrix has to be developed. To date, a vast range of materials has been developed and employed in biodegradable or nonbiodegradable polymer-based drug delivery systems. Nonbiodegradable polymers have been employed in the past; however, their main disadvantage is the need for removal of the subsequent polymer once the drug has depleted, which often leads to surgery. Because of their desirable biocompatible and biodegradable properties, PLGA (see chemical structure illustration) has been widely studied for use as drug delivery vehicles for long-term sustained-release preparations. It capitalizes on the fact that the components of this polymer (lactic and glycolic acids) are metabolites easily handled by the body. Presently, there are several methods for preparing NPs and numerous methods for incorporating drugs into the particles based on the drug of interest, the desired delivery path, and the release profile.

In this laboratory session, a modified version of the SESD method (a general schematic of the NP preparation is provided) has been used for the synthesis of biodegradable PLGA NPs. In this process, nanosize particles of PLGA can be successfully synthesized by pouring the polymeric organic solution into an aqueous phase (surfactant solution/polyvinyl alcohol [PVA] solution; see the chemical structure illustration) with mechanical stirring. In this modified method of the SESD method, PLGA NPs were prepared using solvent systems consisting of two water-miscible organic solvents, such as acetone and ethanol, as the solvent of the polymeric solution instead of using water-immiscible organic solvents (e.g., dichloromethane), and the particles are formed via an emulsification process and then a solvent evaporation process.

In this experiment, we will be using a modified version of the SESD method to prepare biodegradable PLGA NPs.

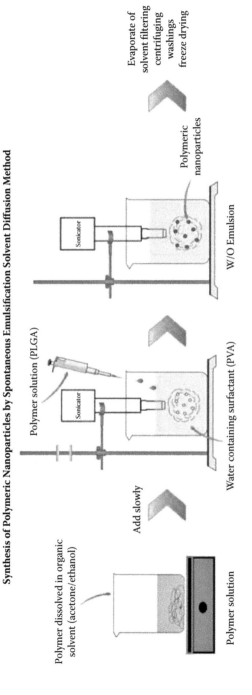

Overall synthesis of nanopolymer by the spontaneous emulsification solvent diffusion method.

poly (lactide-co-glycolide) acid [PLGA]
chemical structure
X = number of units of lactic acid
Y = number of units of glycolic acid

Polyvinyl alcohol (PVA)
chemical structure

Chemical structures of PLGA and PVA.

5.7.3 Key Concepts

1. Synthesis of nanopolymers
2. The SESD method
3. Ultrasonochemistry

5.7.4 Experimental

5.7.4.1 Materials/Reagents

PLGA NPs prepared will depend on the PLGA polymer to which you have access.

Poly(lactide-co-glycolide) acid	PLGA 75:25	$([C_2H_2O_2]_x [C_3H_4O_2]_y)$	x = 75, y = 25
Poly(lactide-co-glycolide) acid	PLGA 50:50	$([C_2H_2O_2]_x [C_3H_4O_2]_y)$	x = 50, y = 50
Poly(lactide-co-glycolide) acid	PLGA 25:75	$([C_2H_2O_2]_x [C_3H_4O_2]_y)$	x = 25, y = 75
Polyvinyl alcohol	2% w/v	$[(C_2H_4O)_n]$	
Absolute ethanol		(CH_3CH_2OH)	
Acetone		(C_3H_6O)	
Milli-Q water			

5.7.4.2 Glassware/Equipment

Clean glassware is essential for this experiment

Measuring cylinder (10 mL)
Beaker (100 mL)
Glass vials (20 mL)
Pasteur pipettes
Micropipette (200–1000 μL)
Micropipette tips (200–1000 μL)
Analytical balance
Hot plate
Sonicator/homogenizer
Kimwipes
Spatula
Magnetic stirrers
Digital camera

5.7.5 Special Safety Precautions

1. Before starting the experiment, look up the Materials Safety Data Sheets for all the chemicals that will be used. All chemicals used should be regarded as irritating and toxic; avoid inhalation.

2. Wear gloves, laboratory coats, and safety eyewear at all times while in the laboratory.

3. Clean glassware is essential for this experiment.

4. Organic solvents are used in this experiment; use great care to avoid inhalation of any organic fumes.

5. Heat is used in this experiment; take care to avoid exposure.

5.7.6 Procedure: Synthesis of PLGA Nanoparticles by a Modified Version of the Spontaneous Emulsification Solvent Diffusion Method

- For each experimental procedure, note any observable changes in the reaction, such as a color change or precipitate formation.

- For each observation, explain what was occurring during the reaction process.

- Take digital photographs of the experimental reactions to be included in your final report.

Part I: Preparation polymer (PLGA) solution

Part II: Preparing the surfactant (PVA) solution

Part III: Preparing PLGA NPs

5.7.6.1 Part I: Preparing Polymer (PLGA) Solution

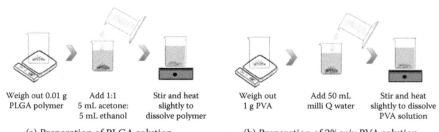

| Weigh out 0.01 g PLGA polymer | Add 1:1 5 mL acetone: 5 mL ethanol | Stir and heat slightly to dissolve polymer | Weigh out 1 g PVA | Add 50 mL milli Q water | Stir and heat slightly to dissolve PVA solution |

(a) Preparation of PLGA solution (b) Preparation of 2% w/v PVA solution

Preparation of PLGA/surfactant solutions.

Part I: Preparing Polymer (PLGA) Solution

Procedure	Observations	Inference/ Comments
Weigh 0.01 g of PLGA polymer ([PLGA 75:25], [PLGA 50: 50], [PLGA 25:75]). *PLGA NPs prepared will depend on the PLGA polymer to which you have access.*		
Add 1:1 of 5 mL of acetone and 5 mL of ethanol organic solvent mixture to the PLGA polymer and fully dissolve the polymer. Heat slightly if polymer does not dissolve easily. A magnetic stirrer can assist in the dissolving polymer.		

5.7.6.2 Part II: Preparing the Surfactant (Polyvinyl Alcohol)

Part II: Preparing the Surfactant (Polyvinyl Alcohol) Solution

Procedure	Observations	Inference/Comments
Add 1 g of polyvinyl alcohol (PVA) to 50 mL of Milli-Q water to produce a 2% w/v solution and fully dissolve PVA. Heat slightly if PVA does not dissolve easily. A magnetic stirrer can assist in dissolution.		

5.7.6.3 Part III: Preparing PLGA Nanoparticles

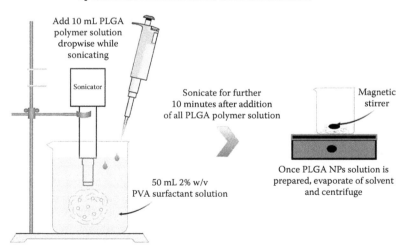

Synthesis of polymeric PLGA nanoparticles by the spontaneous emulsification solvent diffusion method.

Part III: Preparing PLGA Nanoparticles

Procedure	Observations	Inference/ Comments
Sonicate 50 mL of 2% w/v PVA solution using the strongest power of the sonicator or homogenizer.		
To the PVA solution, add the dissolved polymer solution dropwise slowly until all the polymer is gone.		
Once the entire polymer mixture is added, sonicate or homogenize the NP solution for a further 10 minutes.		
Leave the solution to evaporate the solvent by using a magnetic stirrer.		
Once solvent is evaporated, dilute the PLGA NP mixture with Milli-Q water and centrifuge the solution to purify the NP solution.		

5.7.7 Characterization of Polymeric Nanoparticles

Once the polymeric NPs have been synthesized, these samples can be kept for further analysis.

The morphology and shape of the NPs can be analyzed by these techniques:

- FE-SEM
- TEM
- AFM

FURTHER READING MATERIAL

1. Panyam, J.; Labhasetwar, V. Biodegradable nanoparticles for drug and gene delivery to cells and tissue. *Adv. Drug Deliv. Rev.* **2003,** *55* (3), 329–347.
2. Soppimath, K. S.; Aminabhavi, T. M.; Kulkarni, A. R.; Rudzinski, W. E. Biodegradable polymeric nanoparticles as drug delivery devices. *J. Controlled Release* **2001,** *70* (1), 1–20.
3. Bisht, S.; Feldmann, G.; Soni, S.; Ravi, R.; Karikar, C.; Maitra, A.; Maitra, A. Polymeric nanoparticle-encapsulated curcumin ("nanocurcumin"): a novel strategy for human cancer therapy. *J. Nanobiotechnol.* **2007,** *5* (3), 1–18.
4. Niwa, T.; Takeuchi, H.; Hino, T.; Kunou, N.; Kawashima, Y. *In vitro* drug release behavior of D, L-lactide/glycolide copolymer (PLGA) nanospheres with nafarelin acetate prepared by a novel spontaneous emulsification solvent diffusion method. *J. Pharm. Sci.* **1994,** *83* (5), 727–732.

5.8 NANOFORENSICS: FINGERPRINT ANALYSIS STEPPING TOWARD THE NANOWORLD

5.8.1 Aim

The aim is to generate a visible fingerprint mark and analyze it with both optical microscopy and scanning electron microscopy.

5.8.2 Introduction

Popular shows such as *CSI* (*Crime Scene Investigation*) *Las Vegas*, *CSI Miami,* and other police shows have made Locard's principle popular because it is often used to explain the evidence left by a perpetrator at a crime scene. Edmond Locard was a twentieth-century Frenchman who, while working at the first-ever crime laboratory set up in Lyon, France, formulated the famous principle that bears his name. This principle states that the perpetrator's arrival and departure from a crime scene will leave traces such as footprints, fingermarks, and hair fibers from his clothing, as well as broken glass or any tools brought to the crime scene. These act as silent witnesses to the perpetrator's actions and deeds. This principle or theory is almost universally accepted by forensics scientists all over the world and helps to catch the perpetrator of the crime.

One of these important traces left at a crime scene is the fingerprint or fingermark of the perpetrator if he or she was not wearing gloves. It is a series of curved ridges of the folds on the tips of the fingers, and these ridges have distinct patterns. This is a unique item and was first discovered by the Chinese, who used it to sign legal documents. Henry Fauld, while working in a hospital in Japan, published his views about using fingerprints as personal identification, and in 1880, he remarked that it could be an important tool to indentify criminals. Furthermore, fingerprints were also intensely researched by Sir Francis Galton; in 1892, he published his textbook *Finger Prints*. The book showed that a fingerprint has components such as loops, arches, and whorls (refer to the first illustration for this laboratory), that it does not change with age, and most important, that no two prints are identical. Police use of fingerprints started in Argentina, where Dr. Juan Vucetich, who was fascinated by Galton's work, made it into a workable concept. This was followed by the system created by Sir Edward Richard Henry, and it is the same system that was adopted by Scotland Yard. Today, most English-speaking countries, including the United States, use Henry's classification system.

Whorl   Loop   Arch

Main features in a latent fingerprint.

Furthermore, fingerprints are of great importance for indentifying victims of fires, floods, and other disasters (i.e., as was shown in the Asian tsunami of 2004).

There are several different types of fingermark that can be created by the hands; the most common one is the latent fingermark. This is a residue that is essentially an emulsion of sweat and other lipid components. It is about 0.1 μm thick and weighs about 10 μg. Even in each fingerprint, there are several features that need characterization for proper identification (see the second figure for this laboratory). There are other various components in the fingerprint residue; these include water, proteins, lipids, and salts, mostly derived from secretions of the sebaceous, eccrine, and apocrine glands. The residue can also contain cosmetic products as well as food fragments.

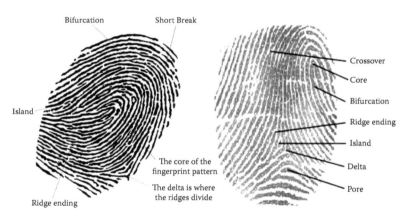

Further classification of smaller features in a fingerprint.

One popular method of making a fingerprint visible at a crime scene is by dusting a very fine fingerprint powder (carbon black, fine charcoal, or aluminum or iron oxides) over the area. This allows the powder to attach to the fingermark residue and thus make the contours of the fingermark visible to the naked eye. However, in some cases, when the surface is a nonporous surface that is not smooth, the traditional method is not efficient; the "superglue" method can be used to reveal the fingerprint. In this case, the substrate holding the suspected fingerprint is allowed to come into contact with a cyanoacrylate, a superglue, which reacts with the materials in the print. The print can then be visualized or dyed to show the fingermark more clearly.

Nanotechnology offers new avenues in the area of visualization of latent fingerprints. Photoluminescent NPs are currently being developed and optimized. Some of these luminescent NPs include functionalized silica (SiO_2) and zinc sulfide (ZnS). QDs (semiconductor nanocrystals) such as CdS and CdSe are also being tested, as their luminous properties are greater than fluorescent dyes presently in use. Furthermore, the size-dependent property of the nanocrystal can be fine-tuned to a narrow emission range and enhance its detection. There is a general consensus by many that nanotechnology will offer better avenues and more sensitive methods for visualization and detection of traces from a perpetrator at a crime scene.

In this lab, you will generate visible fingerprints with a carbon powder to replicate normal fingerprint-dusting technique and create visible fingerprints by the superglue method. These will then be imaged with both optical and electron microscopy and the data compared.

5.8.3 Key Concepts

1. Fingerprint generation

2. Fingerprint analysis and classification

3. Fast enhancement of a fingerprint with carbon toner material

4. Enhancement and identification of fingerprint features by superglue fuming

5.8.4 Experimental

5.8.4.1 Materials/Reagents

Carbon powder (from a laser toner)
Milli-Q water
Superglue

5.8.4.2 Glassware/Equipment

Small aluminum tray
Closed box/evaporation dish
Fume hood
Soft cloth
Wash glass

Continued

Clean glass microscope slides
Superglue
Carbon toner powder
SEM stubs
Small labels
Digital camera
Optical microscope with camera attached
Scanning electron microscope

5.8.5 Special Safety Precautions

- All chemicals used in this experiment should be regarded as irritating and toxic in nature and should be handled with care.

- Lab coats, gloves, and safety glasses should be worn for protection.

- Do not inhale any carbon toner or the superglue.

5.8.6 Procedure

- For each experimental procedure, note any observable changes, such as size, shape, and contours of fingerprints.

- Take digital photographs of the experimental procedure to be included in your final report.

5.8.6.1 Part I: Preparation of Fingerprint Slides

Fingerprint Preparation on Glass Slides

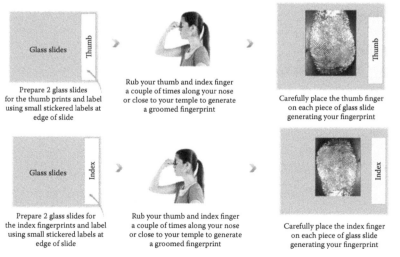

Fingerprint preparation and deposition on glass slides.

Forefinger and Thumb	
Rub your forefinger and thumb a couple of times along your nose or close to your temple to generate a groomed fingerprint and to pick up oil.	
Once you have done this, carefully place the forefinger/thumb on a clean glass slide, generating your fingerprint; label the glass slide.	
Prepare two glass slides for the forefinger prints. **Prepare two slides for the thumbprints.** **Record your observation of the shape and size of the prints.** ***Remember to put gloves back on after using the ungloved finger/thumb to create the print.**	

	Observations
Forefinger	
Thumb	

5.8.6.2 Part II: Preparation of Carbon Toner Fingerprint

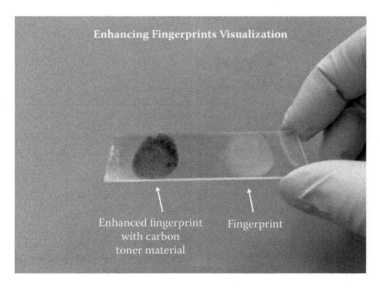

Application of carbon toner material to fingerprint for enhancement.

Prepare two glass slides of fingerprints from carbon toner. Record your observations of the shape and size of the fingerprint.	
Thumb and Forefinger	
For two of the glass slides prepared (**perform in a fume hood**): Place the slides into the carbon toner or dust the carbon toner gently over the slide. Tap any excess of, making sure that none of the carbon toner becomes airborne. Using an optical microscope (preferably attached to a digital camera), analyze the shape and contours of your fingerprint and record the information.	

Carbon Toner	Observations (Eye)	Observations of Features Such as Optical Microscopy
Forefinger		
Thumb		

5.8.6.3 Part III: Preparation for Superglue Fingerprint-Fuming Process

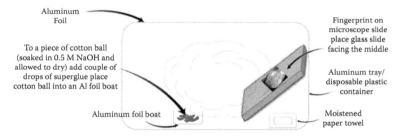

Procedure for superglue fuming to enhance fingerprints.

Prepare two glass slides of fingerprints from superglue. Record your observations of the shape and size of the fingerprints.	
Thumb and Forefinger	
For two of the glass slides prepared: Place the slides into a receptacle and place 2 drops of superglue in the container as shown. Cover the container and leave it for at least 10 minutes (**performed in a fume hood**). After this time, remove the slide and observe the fingerprint mark.	
The example of a fingerprint deposited on a slide is observed with an optical microscope (preferably attached to a digital camera); analyze the shape and contours of your fingerprint and record them.	

Superglue	Observations (Eye)	Observations (Optical Microscope)
Forefinger		
Thumb		

5.8.7 Characterization of Fingerprint Analysis

Once the experiment is complete, the fingerprint slides can be analyzed using the following:

- FE-SEM
- AFM
- Optical microscopy

Plain Finger	Scanning Electron Microscopic Fingerprint Observations
Forefinger	
Thumb	

Carbon Toner	Scanning Electron Microscopic Fingerprint Observations
Forefinger	
Thumb	

Superglue	Scanning Electron Microscopic Fingerprint Observations
Forefinger	
Thumb	

QUESTIONS

1. Based on this laboratory session, what are the general features of your fingerprint? Use the optical microscope and the FE-SEM observations to determine the type of fingerprint you have:

_____

_____

_____

_____

Summary	Fingerprint Observation Characteristics (Type, Ridge, Count, Description)
Forefinger	
Thumb	

Type of fingerprint (loop, arch, whorl or combination):

Forefinger: _____

Thumb: _____

FURTHER READING MATERIAL

1. Boyd A. Comparing fingerprints. http://www.uh.edu/engines/epi2529.htm (accessed Apr 28, 2014).
2. The analysis of crime scenes, cyanoacrylate fuming. http://analyzingacrime-scene.synthasite.com/cyanoacrylate-fuming.php (accessed Apr 28, 2014).
3. Mainguet J. F. Fingerprint, palmprint, pores. http://fingerchip.pagesperso-orange.fr/biometrics/types/fingerprint.htm (accessed Apr 28, 2014).
4. Hong, L.; Jain, A. Classification of fingerprint images. *Proceedings of the Scandinavian Conference on Image Analysis* **1999,** 665–672.
5. Welcome to the California Statewide fingerprint imaging system. Fingerprint pattern types. http://www.sfis.ca.gov/pattern_types.html (accessed Apr 28, 2014).
6. Menzel E. R. Photoluminescence detection of latent fingerprints with quantum dots for time-resolved imaging. *Fingerprint World* **2000,** *26* (101), 119–123.

5.9 SYNTHESIS OF ALGINATE BEADS AND INVESTIGATION OF CITRIC ACID RELEASE FROM A NANOSHELL COATING OF POLYMER

5.9.1 Aim

The aim is to synthesize drug-/dye-loaded alginate (ALG) capsules and to investigate controlled drug/dye release profiles.

5.9.2 Introduction

One of the most exciting and attractive areas of research in drug delivery today is the design and development of nanosystem platforms capable of delivering drugs and medication to the right place, at appropriate times, and at the right dosage *in situ* for an optimal clinical outcome. Traditional drug release by conventional systems (as in the first illustration for this laboratory) can be improved by developing sustained drug release systems (also see the first illustration). Nanoparticulate delivery systems have the potential power to improve drug stability, increase the duration of the therapeutic effect, and minimize drug degradation. Nanoparticles consisting of synthetic biodegradable polymers, natural polymers, lipids, and polysaccharides have been developed and tested over the past decades.

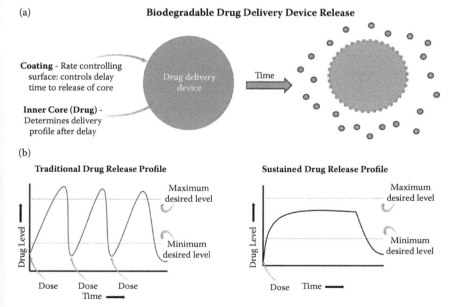

(a) Schematic of drug release system and (b) the traditional and optimal sustained drug release system profiles.

Recently, the idea of using natural and synthetic biodegradable polymers to deliver drugs has provoked great interest. Among them, ALG, chitosan (CS), gelatin, poly-2-hydroxyethyl-methacrylate (pHEMA), and guar gum have shown promising results and are currently exploited for controlling drug release. Biocompatible polymers are the premier choice as these are degraded to molecules that can be handled by the body without serious side effects or long-term issues. An example of such a polymer is CS, a linear polysaccharide macromolecule consisting of glucosamine and N-acetylglucosamine units. It is biocompatible, biodegradable, and nontoxic and is used in the oral drug delivery. Another common biocompatible polymer is pHEMA, which is also used in tissue engineering scaffolds and disposable optical contact lenses.

Micrometer-size particles consisting of synthetic biodegradable polymers, natural biopolymers, lipids, and polysaccharides have been developed and tested over the past decades. A promising natural polymer candidate is ALG, which has been exploited in the pharmaceutical industry for controlling drug release. Alginate is a type of polysaccharide that occurs naturally in all brown algae as a skeletal component of their cell walls. Alginate is also commonly used in food because it is a powerful thickening, stabilizing, and gel-forming agent. It is normally associated with sodium ions as long linear molecules (see the second illustration in this laboratory), and when in contact with calcium ions, there is a strong tendency to cross-link (also see the second illustration), and so the ALG material stiffens. Some foods that may include ALG are ice cream, fruit-filled snacks, salad dressings, pudding, onion rings, and even the pimento strips that are found in the pitted core of green olives.

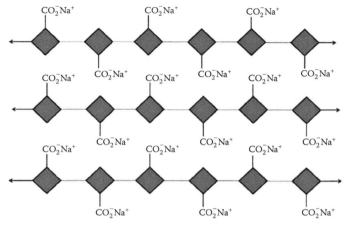

(a) Alginate polymer in NaCl solution (no cross-linking)

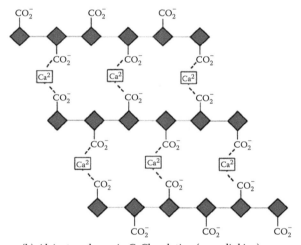

(b) Alginate polymer in $CaCl_2$ solution (cross-linking)

(a) Alginate linear molecules associated with sodium ions and (b) alginate molecules associated with calcium ions leading to cross-linking.

Alginate is biodegradable and biocompatible and has the potential for numerous pharmaceutical and biomedical applications, such as drug delivery systems and cellular encapsulation. Encapsulation of liquids or solids within a membrane is a convenient method to protect these materials or allow their gradual release.

This experiment is a scaled-up version that demonstrates the encapsulation and release of food coloring dye as a visual indicator of capsule formation and how an encapsulated NP loaded with a drug/targeting agent might release its drug load. Diffusion of the food coloring serves as a demonstration of controlled release because it is visible within a few minutes but can take hours to completely leach out.

5.9.3 Key Concepts

1. Drug delivery devices based on polymers

2. Creation of specific-diameter beads of ALG

3. Drug delivery release profiles

4. Use of natural and synthetic biodegradable polymers as coatings for controlling drug release

5.9.4 Experimental

5.9.4.1 Materials/Reagents

Alginic acid sodium salt (ALG) $[(C_6H_8O_6)_n]$	1% w/v
Citric acid $(C_6H_8O_7)$	1% w/v
Calcium chloride $(CaCl_2)$	0.05 M
Poly-2-hydroxyethyl-methacrylate (pHEMA)	1% w/v
Gelatin	1% w/v
Arabic gum (guar gum)	1% w/v
Agar agar	1% w/v
Chitosan	1% w/v
Sodium hydroxide (NaOH)	1 M
Absolute ethanol (CH_3CH_2OH)	
Phenolphthalein	
Milli-Q water	

5.9.4.2 Glassware/Equipment
Clean glassware is required.

Magnetic stirrer	Plastic UV cuvette
Syringe	pH meter
Pasteur pipette	Measuring cylinders (100 mL, 50 mL)

Continued

Beakers (50 mL)	UV spectrometer
Small vials (20 mL)	Spatula
Petri dishes	Balance
Schott bottles (100 mL)	Hot plate
Buchner funnel	Graph paper
Vortex mixer/sonicator	Digital camera
Food coloring dye (yellow, blue, red)	Kimwipes
pH indicator paper	

5.9.5 Special Safety Precautions

- All chemicals used in this experiment should be regarded as irritating and toxic in nature and should be handled with care.

- Lab coats, gloves, and safety glasses should be worn for protection.

- Clean glassware is important for all these experiments.

5.9.6 Procedure: Synthesis and Drug Release Profile of Drug-Loaded Alginate Capsules

- For each experimental procedure, note any observable changes in the reactions, such as color change, change in size, or change in shape.

- For each observation, explain what was occurring during the reaction process.

- Take digital photographs of the experimental reactions to be included in your final report.

Part I: Preparation of solutions and ALG beads

Part IIa: Formation and encapsulation of ALG beaded capsules

Part IIb: Formation of CS-coated ALG beaded capsules

Part IIIa: Encapsulated dye release studies of ALG beaded capsules

Part IIIb: Acid release studies from encapsulated ALG beaded capsules

5.9.6.1 Part I: Preparation of Solutions and Alginate Beads

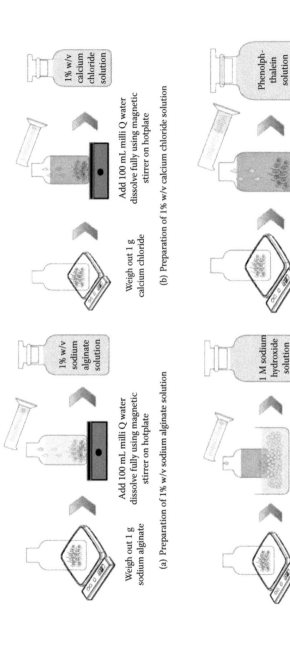

(a) Preparation of 1% w/v sodium alginate solution

Weigh out 1 g sodium alginate → Add 100 mL milli Q water dissolve fully using magnetic stirrer on hotplate → 1% w/v sodium alginate solution

(b) Preparation of 1% w/v calcium chloride solution

Weigh out 1 g calcium chloride → Add 100 mL milli Q water dissolve fully using magnetic stirrer on hotplate → 1% w/v calcium chloride solution

(c) Preparation of 1 M sodium hydroxide solution

Weigh out 6 g sodium hydroxide → Add 50 mL milli Q water and dissolve (in ice mixture since reaction is very vigorous) → 1 M sodium hydroxide solution

(d) Preparation of phenolphthalein solution

Weigh out 0.1 g phenolphthalein → Add 50 mL ethanol and 50 mL milli Q water dissolve fully → Phenolphthalein solution

Preparation of precursor solutions.

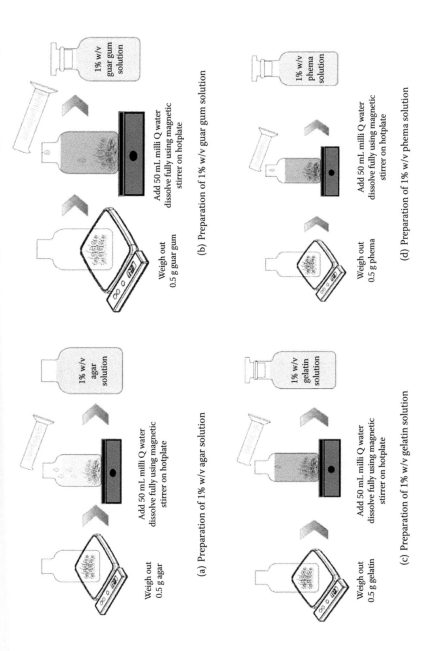

(a) Preparation of 1% w/v agar solution

Weigh out
0.5 g agar

Add 50 mL milli Q water
dissolve fully using magnetic
stirrer on hotplate

1% w/v
agar
solution

(b) Preparation of 1 % w/v guar gum solution

Weigh out
0.5 g guar gum

Add 50 mL milli Q water
dissolve fully using magnetic
stirrer on hotplate

1% w/v
guar gum
solution

(c) Preparation of 1% w/v gelatin solution

Weigh out
0.5 g gelatin

Add 50 mL milli Q water
dissolve fully using magnetic
stirrer on hotplate

1% w/v
gelatin
solution

(d) Preparation of 1% w/v phema solution

Weigh out
0.5 g phema

Add 50 mL milli Q water
dissolve fully using magnetic
stirrer on hotplate

1% w/v
phema
solution

Preparation of polymeric precursor solutions.

Preparation of 1% w/v Chitosan-Citric Acid Solution

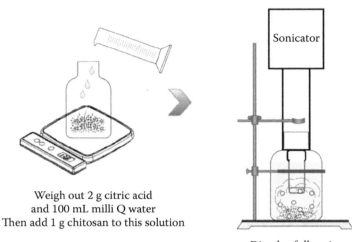

Weigh out 2 g citric acid
and 100 mL milli Q water
Then add 1 g chitosan to this solution

Dissolve fully using
sonicator

Preparation of 1% w/v solutions of chitosan and citric acid by ultrasound.

Preparation of Alginate Beads

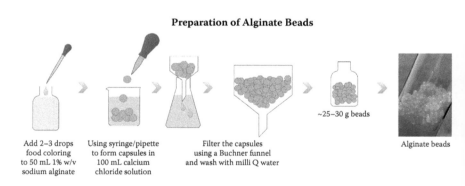

| Add 2–3 drops food coloring to 50 mL 1% w/v sodium alginate | Using syringe/pipette to form capsules in 100 mL calcium chloride solution | Filter the capsules using a Buchner funnel and wash with milli Q water | ~25–30 g beads | Alginate beads |

Preparation of dye-loaded alginate beads.

Part I: Preparation of Solutions and Alginate Beads

Preparing 1% w/v Sodium Alginate Solution (Alginate)		
	Observations	Inferences/Comments
Weigh 0.5 g of alginic acid sodium salt, then add 50 mL Milli-Q water and dissolve fully using a magnetic stirrer if needed.		

Continued

Preparing 1% w/v Calcium Chloride Solution		
	Observations	Inferences/Comments
Weigh 1 g calcium chloride salt, then add 100 mL Milli-Q-water and dissolve fully using a magnetic stirrer if needed.		

Preparation of 1% w/v Chitosan-Citric Acid Solution		
	Observations	Inferences/Comments
Weigh 2 g of citric acid, then add 100 mL Milli-Q water to make a 2% solution.		
Weigh 1 g chitosan and add the 2% citric acid solution; dissolve using the sonicator.		

Preparing 1% w/v Agar Solution		
	Observations	Inferences/Comments
Weigh 0.5 g agar, then add 50 mL Milli-Q water; dissolve fully by heating and using a magnetic stirrer.		

Preparing 1% w/v Guar Gum Solution		
	Observations	Inferences/Comments
Weigh 0.5 g guar gum, then add 50 mL Milli-Q water; dissolve fully using a magnetic stirrer if needed.		

Preparing 1% w/v Gelatin Solution		
	Observations	Inferences/Comments
Weigh 0.5 g gelatin, then add 50 mL Milli-Q water; dissolve fully by heating and using a magnetic stirrer.		

Preparing 1% w/v pHEMA Solution		
	Observations	Inferences/Comments
Weigh 0.5 g pHEMA, then add 50 mL ethanol; dissolve fully using a magnetic stirrer.		

Preparing Phenolphthalein Indicator Solution		
	Observations	Inferences/Comments
Weigh 0.1 g of phenolphthalein, then add 50 mL ethanol and 50 mL of Milli-Q water and dissolve.		

Continued

Preparing 1 M Sodium Hydroxide Solution		
	Observations	Inferences/Comments
Weigh 2 g of sodium hydroxide, then add 50 mL Milli-Q water and dissolve (**in ice mixture because reaction is very vigorous**).		

5.9.6.2 Part IIa: Formation and Encapsulation of Alginate Beaded Capsules

Formation of Chitosan Coated Alginate Beaded Capsules

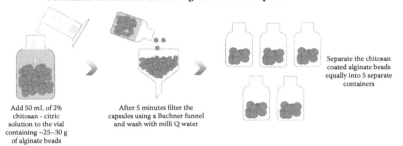

Add 50 mL of 2% chitosan – citric solution to the vial containing ~25–30 g of alginate beads

After 5 minutes filter the capsules using a Buchner funnel and wash with milli Q water

Separate the chitosan coated alginate beads equally into 5 separate containers

Formation of chitosan/citric-acid-coated alginate beads.

Part IIa: Formation and Encapsulation of Alginate Beaded Capsules

	Observations	Inferences/ Comments
To a solution of sodium alginate (50 mL), add a few drops of colored food dye.		
Using a syringe/Pasteur pipette, the colored sodium alginate solution is added dropwise to a beaker of calcium chloride solution (100 mL) to form alginate capsules of similar size.		
Collect the capsules using a Buchner funnel (no paper and no vacuum).		
Wash the alginate beads with Milli-Q water.		
Note: This preparation step is used for further work; make sure plenty of alginate beaded core capsules are produced.		

Separate the alginate beads into 2 separate containers:
    First container: 5 g of alginate beads (alginate/Ca^{2+})
    Second container: 30 g of alginate beads (chitosan shelled alginate beads are chitosan/alginate/Ca^{2+})		

5.9.6.3 Part IIb: Formation of Chitosan-Coated Alginate Beaded Capsules

Part IIb: Formation of Chitosan-Coated Alginate Beaded Capsules

	Observations	Inferences/Comments
To the second container containing 30 g of alginate capsules, add the 2% chitosan-citric acid solution (50 mL).		
After 5 minutes, the chitosan-citric acid solution is removed and capsules collected using a Buchner funnel.		
The collected beads are washed with Milli-Q water.		
Separate the beads into 6 vials containing 5 g of alginate beads each. One vial will be used in Part IV, and 5 vials will be used in Part V.		

5.9.6.4 Part IIIa: Encapsulated Dye Release Studies of Alginate Beaded Capsules

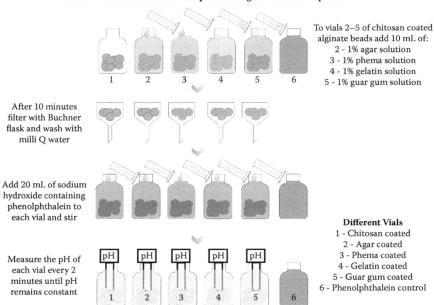

Release of Acid Studies from Encapsulated Alginate Beaded Capsules

To vials 2–5 of chitosan coated alginate beads add 10 mL of:
2 - 1% agar solution
3 - 1% phema solution
4 - 1% gelatin solution
5 - 1% guar gum solution

After 10 minutes filter with Buchner flask and wash with milli Q water

Add 20 mL of sodium hydroxide containing phenolphthalein to each vial and stir

Measure the pH of each vial every 2 minutes until pH remains constant

Different Vials
1 - Chitosan coated
2 - Agar coated
3 - Phema coated
4 - Gelatin coated
5 - Guar gum coated
6 - Phenolphthalein control

Schematic procedure for the acid release studies from the coated alginate beads.

Part IIIa: Encapsulated Dye Release Studies of Alginate Beaded Capsules
Dye release is compared between the two alginate-beaded portions using Milli-Q water.
1: Nonshelled alginate beads (5 g)
2: Chitosan shelled alginate beads (5 g)

	Observations	Inferences/Comments
To each vial (nonshelled and chitosan shelled alginate beads), add Milli-Q water (50 mL).		
For both vials: Measure UV absorption of 1 mL solution using UV absorption spectroscopy every 15 minutes for 2 hours to determine the concentration of dye release. *Note:* Diffusion of the coloring dye serves as demonstration of controlled release because it is visible in solution. *For each, note any observable changes in the reactions, such as color change, change in size, change in shape.*		

Analysis

1. Plot time versus concentration to determine the rate of dye release for each vial.
2. Compare how long the dye is released from the 2 vials.

1. Plot Time versus Absorption Graph to Determine Concentration of Dye Release Vial 1: Nonshelled Alginate Beads			
Time	UV Absorption	Concentration	Observations
0 minutes			
15 minutes			
30 minutes			
45 minutes			
60 minutes			
75 minutes			
90 minutes			
105 minutes			
120 minutes			

Vial 2: Chitosan Shelled Alginate Beads			
Time	UV Absorption	Concentration	Observations
0 minutes			
15 minutes			
30 minutes			
45 minutes			
60 minutes			

Continued

Vial 2: Chitosan Shelled Alginate Beads			
Time	UV Absorption	Concentration	Observations
75 minutes			
90 minutes			
105 minutes			
120 minutes			

5.9.6.5 Part IIIb: Acid Release Studies from Encapsulated Alginate Beaded Capsules

Part IIIb: Acid Release Studies from Encapsulated Alginate Beaded Capsules

Acid release profiles are compared between the 5 alginate beaded portions using dilute sodium hydroxide (1 M) solution. The chitosan-coated alginate beads from part III are used here.		
Vial 1	Chitosan-coated alginate beads	
Vial 2	Agar shell with chitosan-coated alginate bead core	
Vial 3	pHEMA shell with chitosan-coated alginate bead core	
Vial 4	Gelatin shell with chitosan-coated alginate bead core	
Vial 5	Guar gum shell with chitosan-coated alginate bead core	

	Observations	Inferences/ Comments
To each vial of chitosan-coated alginate beads, add Vial 2: 10 mL 1% agar solution Vial 3: 10 mL 1% pHEMA solution Vial 4: 10 mL 1% gelatin solution Vial 5: 10 mL 1% guar gum solution		
After 10 minutes, collect the beads of each vial with a Buchner funnel.		
Wash the beads with Milli-Q water.		
Put the beads of each vial into 5 separate labeled vials.		

Continued

	Observations	Inferences/ Comments
To 120 mL of 1 M sodium hydroxide, add a few drops of phenolphthalein and stir. Check pH using a pH meter.		
Add 20 mL of the 1 M NaOH to the 5 vials of alginate beads.		
Add 20 mL of 1 M NaOH to an empty vial as a control.		
Measure the pH of each vial every 2 minutes until the pH remains constant and record. *For each, note any observable changes in the reactions, such as color change, change in size, change in shape.*		

Analysis

1. Plot pH versus time for each vial.
2. Compare the graphs and determine which coating allows greatest release.

Vial 1: Chitosan Coated		
Time (minutes)	**pH**	**Observation**

Continued

Vial 1: Chitosan Coated		
Time (minutes)	pH	Observation

Vial 2: Agar Coated		
Time	pH	Observation

Vial 3: pHEMA Coated		
Time	pH	Observation

Continued

Time	pH	Observation

Vial 4: Gelatin Coated		
Time	pH	Observation

Continued

Vial 4: Gelatin Coated		
Time	pH	Observation

Vial 5: Guar Gum Coated		
Time	pH	Observation

FURTHER READING MATERIAL

1. Pignolet, L. H.; Waldman, A. S.; Schechinger, L.; Govindarajoo, G.; Nowick, J. S.; Labuza, T. The alginate demonstration: polymers, food science, and ion exchange. *J. Chem. Educ.* **1998,** *75* (11), 1430.
2. Bagaria, H. G.; Dean, M. R.; Nichol, C. A.; Wong, M. S. Self-assembly and nanotechnology: real-time, hands-on, and safe experiments for K-12 students. *J. Chem. Educ.* **2011,** *88* (5), 609–614.

5.10 SUPERHYDROPHOBICITY AND SELF-CLEANING EFFECT OF A SURFACE

5.10.1 Aim

The aim is to understand the superhydrophobic nature and self-cleaning effect of lotus leaf or other superhydrophobic plant material.

5.10.2 Introduction

The *lotus effect* refers to the very high water repellence (superhydrophobicity) exhibited by the leaves of the lotus flower (*Nelumbo nucifera*). Dirt particles picked up by water droplets because of the complex micro- and nanoscopic architecture of the leaves' surface minimizes adhesion and allows the leaves to be clean. The hydrophobicity of a surface is related to its contact angle. The higher the contact angle, the higher the hydrophobicity of a surface will be. Surfaces with a contact angle less than 90° are considered generally hydrophilic and those with an angle greater than 90° as hydrophobic. Surfaces with an even greater contact angle of more than 150° are considered superhydrophobic (see the first illustration for this laboratory). Some plant surfaces show contact angles up to 160°, meaning that only 2–3% of a leaf's surface is in contact with the water droplet.

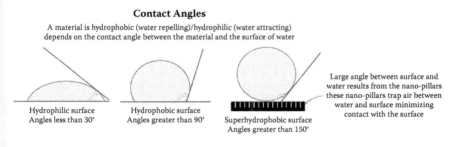

Contact Angles

A material is hydrophobic (water repelling)/hydrophilic (water attracting) depends on the contact angle between the material and the surface of water

Hydrophilic surface
Angles less than 30°

Hydrophobic surface
Angles greater than 90°

Superhydrophobic surface
Angles greater than 150°

Large angle between surface and water results from the nano-pillars these nano-pillars trap air between water and surface minimizing contact with the surface

Contact angle of different surfaces.

Because of their high surface tension, water droplets tend to minimize their surface in an effort to achieve a spherical shape. On contact with a surface, adhesion forces result in wetting of the surface. Either complete or incomplete wetting may occur depending on the structure of the surface and the fluid tension of the droplet. The cause of self-cleaning properties is the hydrophobic water-repellent double structure of the surface. This enables the contact area and the adhesion force between surface and droplet to be significantly reduced, resulting in self-cleaning (see the second illustration for this laboratory).

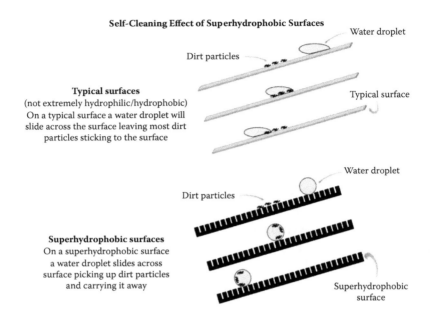

Self-cleaning effect of superhydrophobic surfaces.

Studies have shown that, in the case of *Nelumba nucifera* (sacred lotus leaf), there is interplay between the micrometer-size surface structures and the nano-size wax crystal formations. This interplay effectively increases the surface roughness and makes the leaf surface more water repellant or hydrophobic. So, when water comes into contact with the leaf surface, the water drops do not spread but instead form beads. The beads easily roll off the leaf and in the process remove foreign particles from its surface, thus giving the lotus leaf the added property of self-cleaning.

The wetting of a solid material by water is an important concept and has wide practical applications in both nature and industry. Since the discovery of the lotus effect by Barthlott and Neinhuis in 1997, there has been a continuing search for other similar surfaces in nature. Indeed, taro leaves, Indian watercress, and even the petals of several flowers have been found to display surface structures/features at both the micrometer and nanometer scales that have similar characteristics to the lotus leaves. Recently, the Murdoch University Nanotechnology Research Group (MANRG) has investigated the nano-/microstructures on the leaves of some Australian indigenous eucalyptus plants, in particular one called the mottlecah (see the a–c parts of the third illustration). This arid plant's cuticular membrane displays a remarkable ability to repel water and self-clean.

Other eucalyptus plants exhibit the same self-cleaning properties as the mottlecah with respect to impurities. Removal of particles such as carbon black toner can be performed effectively simply by wetting the leaf (see the d part of the third figure for this lab).

Picture of (a) the mottlecah plant, (b) colored dye water drops sitting on the leaf's surface, (c) the scanning electron image of the nanopillars, and (d) the self-cleaning effect of water droplets collecting carbon black toner material sprayed previously on another eucalyptus leaf.

5.10.3 Key Concepts

1. Demonstration of self-cleaning properties of surfaces
2. Creation of superhydrophobic surfaces
3. Concept of surface tension
4. Concept of contact angle of water with hydrophilic and hydrophobic surfaces

5.10.4 Experimental

5.10.4.1 Materials/Reagents

Milli-Q water
Chloroform ($CHCl_3$)
Black ink toner
Food coloring dye (red, orange, green)

Continued

Glue
Superhydrophobic leaves
Hydrophilic leaves

5.10.4.2 Glassware/Equipment

Microscope slides
Water spray bottle
Beakers
Tweezers
Pasteur pipettes
Micropipettes (10–200 μL)
Micropipette tips (10–200 μL)
Scalpel/scissors
Kimwipes
Fume hood

5.10.5 Special Safety Precautions

- Before starting the experiment, look up the Materials Safety Data Sheets for all the chemicals that will be used. All chemicals used should be regarded as irritating and toxic; avoid inhalation.

- Wear gloves, laboratory coats, and safety eyewear at all times while in the laboratory.

- Special care needs to be taken when handling chloroform; wear gloves and work in a fume hood.

5.10.6 Procedure: Superhydrophobicity and Self-Cleaning Effect of a Surface

Part Ia: Leaf preparation on microscope slides—hydrophilic leaves

Part Ib: Leaf preparation on microscope slides—superhydrophobic leaves

Part IIa: Contact angle estimation of hydrophilic leaves using digital pictures

Part IIb: Contact angle estimation of superhydrophobic leaves using digital pictures

Part IIIa: Self-replicating properties of leaf wax, hydrophilic leaves—contact angle estimation using digital pictures

Part IIIb: Self-replicating properties of leaf wax, superhydrophobic leaves—contact angle estimation using digital pictures

Part IVa: Self-cleaning experiment using carbon black toner—hydrophilic whole leaf

Part IVb: Self-cleaning experiment using carbon black toner—superhydrophobic whole leaf

Part Va: Self-cleaning experiment using carbon black toner, hydrophilic leaves—microscope slides

Part Vb: Self-cleaning experiment using carbon black toner, superhydrophobic leaves—microscope slides

5.10.6.1 Part Ia: Leaf Preparation on Microscope Slides—Hydrophilic Leaves

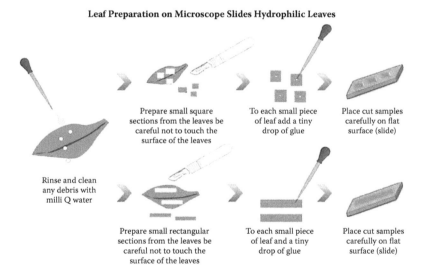

Leaf preparation on microscope slides: hydrophilic leaves.

Part Ia: Leaf Preparation on Microscope Slides—Hydrophilic Leaves

	Observations
Rinse and clean debris off leaves (at least 2 leaves) with Milli-Q water.	
Using a scalpel, cut **small square sections** (same size) from the leaves (be careful not to touch the surface of the leaves).	

Continued

	Observations
Using a scalpel, cut **small rectangular sections** (same size) from leaves (be careful not to touch the surface of the leaves).	
On the small square and rectangular pieces, add a tiny drop of glue on one side of the pieces (be careful not to smear glue on both sides of pieces).	
Prepare a microscope slide with 3 small square sections.	
Prepare a microscope slide with a small rectangular section.	

5.10.6.2 Part Ib: Leaf Preparation on Microscope Slides—Superhydrophobic Leaves

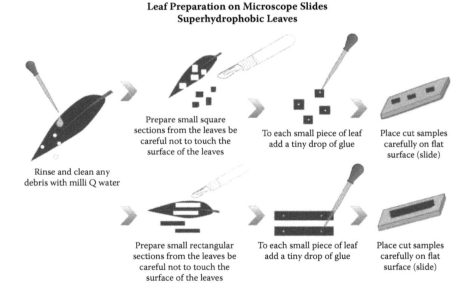

Leaf Preparation on Microscope Slides Superhydrophobic Leaves

Rinse and clean any debris with milli Q water

Prepare small square sections from the leaves be careful not to touch the surface of the leaves

To each small piece of leaf add a tiny drop of glue

Place cut samples carefully on flat surface (slide)

Prepare small rectangular sections from the leaves be careful not to touch the surface of the leaves

To each small piece of leaf add a tiny drop of glue

Place cut samples carefully on flat surface (slide)

Leaf preparation on microscope slides: superhydrophobic leaves.

Part Ib: Leaf Preparation on Microscope Slides—Superhydrophobic Leaves

	Observations
Rinse and clean debris off leaves (at least 2 leaves) with Milli-Q water.	
Using a scalpel, cut **small square sections** (same size) from the leaves (be careful not to touch the surface of the leaves).	

Continued

	Observations
Using a scalpel, cut **small rectangular sections** (same size) from leaves (be careful not to touch the surface of the leaves).	
On the small square and rectangular pieces, add a tiny drop of glue on one side of the pieces (be careful not to smear glue on both sides of pieces).	
Prepare a microscope slide with 3 small square sections.	
Prepare a microscope slide with a small rectangular section.	

5.10.6.3 Part IIa: Contact Angle Estimation of Hydrophilic Leaves Using Digital Pictures

Contact Angle Estimation Using Digital Pictures

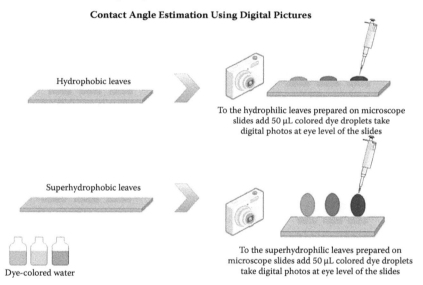

Hydrophobic leaves

To the hydrophilic leaves prepared on microscope slides add 50 µL colored dye droplets take digital photos at eye level of the slides

Superhydrophobic leaves

To the superhydrophilic leaves prepared on microscope slides add 50 µL colored dye droplets take digital photos at eye level of the slides

Dye-colored water

Take multiple digital pictures of each leaf slides to analyze later

Contact angle estimation using digital pictures.

Part IIa: Contact Angle Estimation of Hydrophilic Leaves Using Digital Pictures

	Observations
Take the prepared microscope slide with the 3 small square sections and add 50 µL of dye colored water (red, green, and orange) to each square leaf piece. Be careful that the droplets do not roll onto the other leaf pieces.	

Continued

	Observations
Take multiple digital pictures of the slide using a digital camera. Make sure to take the pictures of the slide at eye level.	
Estimate the contact angle of the water droplets for each leaf piece.	

5.10.6.4 Part IIb: Contact Angle Estimation of Superhydrophobic Leaves Using Digital Pictures

Part IIb: Contact Angle Estimation of Superhydrophobic Leaves Using Digital Pictures

	Observations
Take the prepared microscope slide with the 3 small square sections and add 50 µL of dye colored water (red, green, and orange) to each square leaf piece.	
Take multiple digital pictures of the slide using a digital camera. Make sure to take the pictures of the slide at eye level.	
Estimate the contact angle of the water droplets for each leaf piece.	

5.10.6.5 Part IIIa: Self-Replicating Properties of Leaf Wax, Hydrophilic Leaves—Contact Angle Estimation Using Digital Pictures

Hydrophilic Leaves
Self-replicating properties of leaf wax
Contact angle estimation using digital pictures

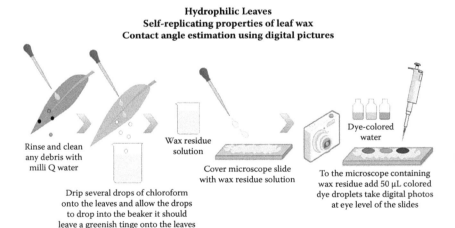

Rinse and clean any debris with milli Q water

Drip several drops of chloroform onto the leaves and allow the drops to drop into the beaker it should leave a greenish tinge onto the leaves

Wax residue solution

Cover microscope slide with wax residue solution

Dye-colored water

To the microscope containing wax residue add 50 µL colored dye droplets take digital photos at eye level of the slides

Take multiple digital pictures of each wax slides to estimate contact angles compare these to original leaf contact angle measurements

Self-replicating properties of leaf wax (hydrophilic leaves) and contact angle estimation using digital pictures.

Part IIIa: Self-Replicating Properties of Leaf Wax, Hydrophilic Leaves: Contact Angle Estimation Using Digital Pictures

	Observations
Rinse and clean debris off leaves (at least 3 leaves) with Milli-Q water.	
To each leaf, drip chloroform onto the leaf and allow the drops of chloroform (containing wax) to drop into a beaker; this should leave a greenish tinge on the leaves.	
Cover a microscope slide completely with the wax residue solution collected; allow the residue to dry. Be careful not to smudge the slide.	
To the microscope slide covered with wax residue, add 50 μL of colored dye water droplets (red, green, orange).	
Take multiple digital pictures of the slide using a digital camera. Make sure to take the pictures of the slide at eye level.	
Estimate the contact angle of the water droplets for each water droplet on the microscope slides.	
Compare the estimated wax residue contact angles with the leaf's estimated contact angles from part II.	

5.10.6.6 Part IIIb: Self-Replicating Properties of Leaf Wax, Superhydrophobic Leaves—Contact Angle Estimation Using Digital Pictures

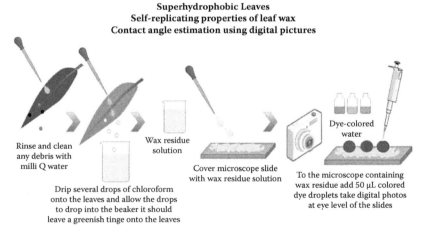

Superhydrophobic Leaves
Self-replicating properties of leaf wax
Contact angle estimation using digital pictures

Rinse and clean any debris with milli Q water

Drip several drops of chloroform onto the leaves and allow the drops to drop into the beaker it should leave a greenish tinge onto the leaves

Wax residue solution

Cover microscope slide with wax residue solution

Dye-colored water

To the microscope containing wax residue add 50 μL colored dye droplets take digital photos at eye level of the slides

Take multiple digital pictures of each wax slides to estimate contact angles compare these to original leaf contact angle measurements

Self-replicating properties of leaf wax (superhydrophobic leaves) and contact angle estimation using digital pictures.

Part III (b): Self-Replicating Properties of Leaf Wax, Superhydrophobic Leaves—
Contact Angle Estimation Using Digital Pictures

	Observations
Rinse and clean debris off leaves (at least 3 leaves) with Milli-Q water.	
To each leaf, drip chloroform onto the leaf and allow the drops of chloroform (containing wax) to drop into a beaker.	
Cover a microscope slide completely with the wax residue solution collected; allow the residue to dry. Be careful not to smudge the slide.	
To the microscope slide covered with wax residue, add 50 μL of colored dye water droplets (red, green, orange).	
Take multiple digital pictures of the slide using a digital camera. Make sure to take the pictures of the slide at eye level.	
Estimate the contact angle of the water droplets for each water droplet on the microscope slides.	
Compare the estimated wax residue contact angles with the leaf's estimated contact angles from part II.	

5.10.6.7 Part IVa: Self-Cleaning Experiment Using Carbon Black Toner—Hydrophilic Whole Leaf

Hydrophilic Leaves
Self-cleaning experiment using black carbon toner

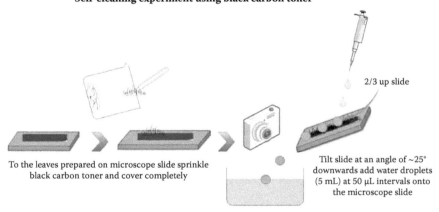

2/3 up slide

To the leaves prepared on microscope slide sprinkle black carbon toner and cover completely

Tilt slide at an angle of ~25° downwards add water droplets (5 mL) at 50 μL intervals onto the microscope slide

Film the motion of the water droplets over the surface of the leaf using a digital camera
Observe the interaction between the water droplets and the carbon black toner
Compare the hydrophilic leaf surface to the superhydrophobic leaf surface

Hydrophilic leaf that sticks to microscopic slide (see Figure 5.10.6.7 Part IVa) and its self-cleaning procedure using carbon black toner.

Part IVa: Self-Cleaning Experiment Using Carbon Black Toner—Whole Hydrophilic Leaf

	Observations
Rinse and clean debris off a leaf with Milli-Q water.	
Dip the leaf into a container of black ink toner powder and coat the leaf completely.	
Film the following with a digital camera: Spray the leaf with water to remove ink toner. Note how many sprays are needed to completely remove ink toner.	
Extract the sequential frames to show the self-cleaning effect of the leaf. Observe the interaction between the water droplets and the carbon black toner.	
Compare the hydrophilic leaf surface after the water spray to the superhydrophobic leaf surface.	

5.10.6.8 Part IVb: Self-Cleaning Experiment Using Carbon Black Toner—Superhydrophobic Whole Leaf

Superhydrophobic Leaves
Self-cleaning experiment using black carbon toner

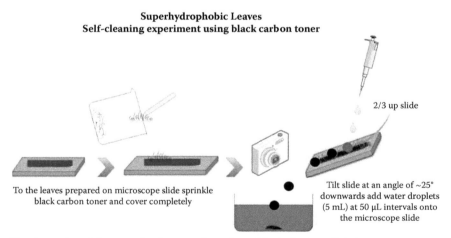

2/3 up slide

To the leaves prepared on microscope slide sprinkle black carbon toner and cover completely

Tilt slide at an angle of ~25° downwards add water droplets (5 mL) at 50 μL intervals onto the microscope slide

Film the motion of the water droplets over the surface of the leaf using a digital camera
Observe the interaction between the water droplets and the carbon black toner
Compare the hydrophilic leaf surface to the superhydrophobic leaf surface

Superhydrophobic leaf that sticks to microscope slide (see Figure 5.10.6.8 Part IVb) and its self-cleaning procedure using carbon black toner.

Part IVb: Self-Cleaning Experiment Using Carbon Black Toner—Whole Superhydrophobic Leaf

	Observations
Rinse and clean debris off a leaf with Milli-Q water.	
Dip the leaf into a container of black ink toner and coat the leaf completely.	
Film the following with a digital camera: Spray the leaf with water to remove ink toner.	
Note how many sprays are needed to completely remove ink toner.	
Extract the sequential frames to show the self-cleaning effect of the leaf. Observe the interaction between the water droplets and the carbon black toner.	
Compare the superhydrophobic leaf surface after the water spray to the hydrophilic leaf surface.	

5.10.6.9 Part Va: Self-Cleaning Experiment Using Carbon Black Toner, Hydrophilic Leaves—Microscope Slides

Hydrophilic Leaves
Self-cleaning experiment using black carbon toner

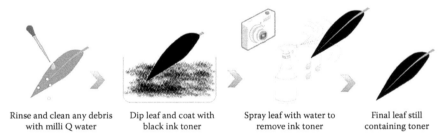

| Rinse and clean any debris with milli Q water | Dip leaf and coat with black ink toner | Spray leaf with water to remove ink toner | Final leaf still containing toner |

Film the motion of the water droplets over the surface of the leaf using a digital camera
Observe the interaction between the water droplets and the carbon black toner
Compare the hydrophilic leaf surface to the superhydrophobic leaf surface

Hydrophilic leaves and their self-cleaning procedure using carbon black toner.

Part Va: Self-Cleaning Experiment Using Carbon Black Toner, Hydrophilic Leaves—Microscope Slides

	Observations
To the prepared microscope slide with the small rectangular leaf piece (prepared in part I), sprinkle black carbon toner and cover it completely.	

Continued

	Observations
Film the following with a digital camera: Tilt the slide at an angle of ~ 25° downward. Add water droplets at 50-μL intervals for 5 mL.	
Extract the sequential frames to show the self-cleaning effect of the leaf. Observe the interaction between the water droplets and the carbon black toner.	
Compare the leaf surface to the original leaf. Compare the hydrophilic leaf surface after water treatment to the superhydrophobic leaf surface.	

5.10.6.10 Part Vb: Self-Cleaning Experiment Using Carbon Black Toner, Superhydrophobic Leaves—Microscope Slides

Superhydrophobic Leaves
Self-cleaning experiment using black carbon toner

| Rinse and clean any debris with milli Q water | Dip leaf and coat with black ink toner | Spray leaf with water to remove ink toner | Final leaf with no more toner |

Film the motion of the water droplets over the surface of the leaf using a digital camera
Observe the interaction between the water droplets and the carbon black toner
Compare the hydrophilic leaf surface to the superhydrophobic leaf surface

Superhydrophobic leaves and their self-cleaning procedure using carbon black toner.

Part Vb: Self-Cleaning Experiment Using Carbon Black Toner, Superhydrophobic Leaves—Microscope Slides

	Observations
To the prepared microscope slide with the small rectangular leaf piece (prepared in part I), sprinkle black carbon toner and cover it completely.	
Film the following with a digital camera: Tilt the slide at an angle of ~ 25° downward. Add water droplets at 50-μL intervals for 5 mL.	
Extract the sequential frames to show the self-cleaning effect of the leaf. Observe the interaction between the water droplets and the carbon black toner.	
Compare the leaf surface to the original leaf. Compare the superhydrophobic leaf surface after water treatment to the hydrophilic leaf surface.	

5.10.7 Contact Angle Estimation

Contact Angle Estimation

		Contact Angle Estimation
Prepared leaf	Hydrophilic leaves	
	Superhydrophobic leaves	
Wax residue	Hydrophilic leaves	
	Superhydrophobic leaves	

FURTHER READING MATERIAL

1. Barthlott, W.; Neinhuis, C. Purity of the sacred lotus, or escape from contamination in biological surfaces. *Planta* **1997,** *202* (1), 1–8.
2. Bhushan, B. Biomimetics: lessons from nature—an overview. *Phil. Trans. R. Soc. A* **2009,** *367* (1893), 1445–1486.
3. Brooker, M. I. H.; Kleinig, D. A. *Field Guide to Eucalypts. Volume 1. Southeastern Australia*; Inkata Press: Melbourne, 1983.
4. Zhang, X.; Shi, F.; Niu, J.; Jiang, Y.; Wang, Z. Superhydrophobic surfaces: from structural control to functional application. *J. Matter. Chem.* **2008,** *18* (6), 621–633.
5. Poinern, G. E. J.; Le, X. T.; Fawcett, D. Superhydrophobic nature of nanostructures on an indigenous Australian eucalyptus plant and its potential application. *Nanotechnol. Sci. Appl.* **2011,** *4,* 113–121.

5.11 SAMPLE ANALYSIS USING SCANNING ELECTRON MICROSCOPY

5.11.1 Aim

The aim is to gain hands-on skills and capability in preparing samples for SEM as well as valuable experience regarding the workings of a scanning electron microscope.

5.11.2 Introduction

The limitations of optical microscopy steered researchers to look for better microscopes, and the electron could be used as a fine probe because of its much smaller wavelength than that of light. The development of this system was only possible with advancements made in areas of electromagnetic lenses to allow for the fine focusing of those high-energy electrons onto the sample. The SEM uses a focused beam of high-energy electrons to image solid samples with very high resolution, and it is better than an ordinary optical microscope. The electron beam interaction gives rise to several phenomena (see the first figure for this laboratory). These phenomena give rise to radiations that can be used to gain much structural and compositional information about the sample under investigation, namely, topography, external morphology, chemical composition, and crystalline phase. Of note, the first commercial SEM was developed by Professor Sir Charles Oatley and his group at Cambridge University. It was then marketed in 1965 by the Cambridge Instrument Company and was known as the Stereoscan.

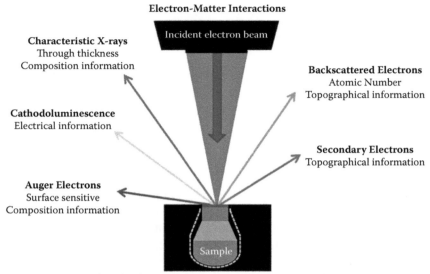

When the electron beam hits a sample it interacts with the atoms in that sample, electrons and x-rays can be the results of these interactions to give information about the sample

Schematic of electron-matter interactions as the SEM sample is investigated under a focused beam of electrons.

Today, the SEM is routinely used instrument in academia and in industry for the examination of the fine detail of a variety of specimens, which may be mineral solids or biological or powdered samples. Some of the samples that can be prepared during the nanotechnology laboratory course are listed in the table below. The modern SEM captures digital images that can be sent electronically anywhere in the world and analyzed within seconds. Given a cost differential compared to an optical microscope, one must ask why SEMs are necessary. Simply, the SEM has much better resolution and array of information and a greater depth of focus than an optical microscope.

Types of Samples Prepared during Laboratory Course

Type of Samples to be Analyzed	Samples from the Labs Performed
Powdered samples > 1 μm	ZnO nanorods
Powdered samples < 1 μm	Raw carbon soot
Solid samples	Fingerprint slides, plant leaves slides

Continued

Type of Samples to be Analyzed	Samples from the Labs Performed
Liquid samples	Au NPs, Ag NPs, ZnS NPs, carbon soot, PLGA NPs, bimetallic NPs (Fe@Ag, Fe@Au)
Biological samples	To be collected before laboratory session: Butterfly wings, moths, colored beetles, common leaves, and superhydrophobic leaves

In this lab, there is the opportunity to image several types of samples that you have synthesized from the previous laboratory sessions as well as explore some nano-/microstructures from the natural world.

5.11.3 Key Concepts

1. Basic operation of an SEM

2. Electron specimen interactions in SEM

3. Sample preparation for SEM investigation

4. Investigation of nanostructures (biological specimen and synthetic materials)

5.11.4 Experimental

5.11.4.1 Materials/Reagents

Highly ordered pyrolytic graphite (HOPG)
Mica
Nitrogen gas
Milli-Q water
SEM samples to be analyzed

5.11.4.2 Glassware/Equipment

SEM stubs	Gloves
SEM carbon or copper tape	Fine-point permanent marker
SEM sample box	Vacuum oven
SEM tweezers	Double-sided adhesive tape
SEM carbon glue	Ultrasonic bath
Scissors	Beaker (10 mL)
Pasteur pipette	Digital camera
Kimwipes	Sputter coater (Au, Pt, C)
Capillary tubes for SEM glue	Micropipettes (10–200 μL)
Spatula	Micropipette tips (10–200 μL)
Tweezers	Fume hood

5.11.5 Special Safety Precautions

1. Before starting the experiment, look up the Materials Safety Data Sheets for all the chemicals that will be used.

2. Wear gloves, laboratory coats, and safety eyewear at all times while in the laboratory.

3. Use the fume hood if handling fine powders.

5.11.6 Procedure: Sample Analysis Using Scanning Electron Microscopy

Prepare SEM samples for analysis with an SEM.

1. Take a photograph of the sample before mounting and after mounting on the SEM stub to include in your laboratory report.

2. Mount the conductive copper/carbon SEM tape onto the SEM stub.

3. Place a mark on the top of the SEM stub; it can be used as a guide to where your sample is under the microscope. Write under the stub a code for the sample; this will help to identify what sample you have for analysis later (see the a portion of the second figure for this laboratory).

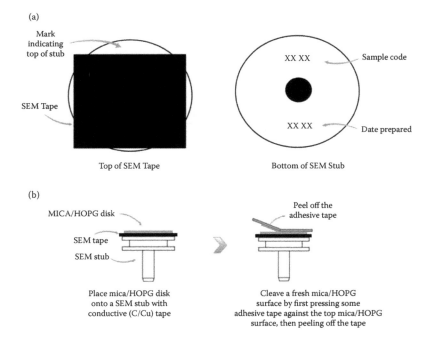

General preparation steps for SEM stub prior to analysis.

4. If analyzing liquid samples, you will need to mount a mica/HOPG disk onto the stub; press double-sided adhesive to the top of mica/HOPG surface, then peel off the tape to obtain a clean surface (see the b portion of the second figure).

5. Mount the sample carefully onto the carbon/copper tape.

   • The type of sample to be analyzed will determine how it is prepared and mounted to the SEM stub. A general description of how to mount the different types of samples follows.

6. Once the sample is mounted, record the sample in detail using the preparation sheet included.

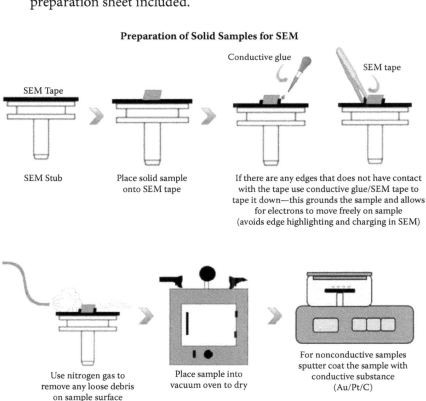

Preparation of Solid Samples for SEM

SEM Tape

SEM Stub

Place solid sample onto SEM tape

Conductive glue

SEM tape

If there are any edges that does not have contact with the tape use conductive glue/SEM tape to tape it down—this grounds the sample and allows for electrons to move freely on sample (avoids edge highlighting and charging in SEM)

Use nitrogen gas to remove any loose debris on sample surface

Place sample into vacuum oven to dry

For nonconductive samples sputter coat the sample with conductive substance (Au/Pt/C)

Sample of an SEM sample preparation record sheet for efficient archiving.

7. Allow the sample to dry using a vacuum oven.

8. Use nitrogen gas to remove any debris on the sample surface, especially powdered samples.

9. Nonconductive samples need to be coated with a conductive substance (C/Au/Pt).

   • **Ensure the sample is stable and dry and no loose particles are present.** The SEM is under high vacuum, so any loose, wet particles can cause severe damage to the SEM.

   • Please record the sample description with as much detail as possible to help with sample analysis later.

   • The type of sample to be analyzed will determine how it is prepared and mounted to the SEM stub.

5.11.6.1 Solid Samples

If the sample is a solid, mount the sample on the carbon/copper SEM tape. The sample needs to be electrically connected to the sample holder/SEM stub to prevent the electron beam from "charging" the sample and distorting the image. Use conductive glue/tape to ensure that any edges that do not have contact with the SEM tape are electrically connected. This is important to help "ground" the sample on the tape and avoid charging and edge highlighting effects. Ensure solid sample is not loose (use conductive tape to ensure the sample is affixed to the SEM stub) or contain any debris (use nitrogen gas to remove any loose debris from the sample surface). If the solid sample is nonconductive, it will need to be coated with a conductive material (i.e., Au/Pt or C). The illustration is of a general sample preparation procedure for solid samples.

SEM Preparation of Biological Samples

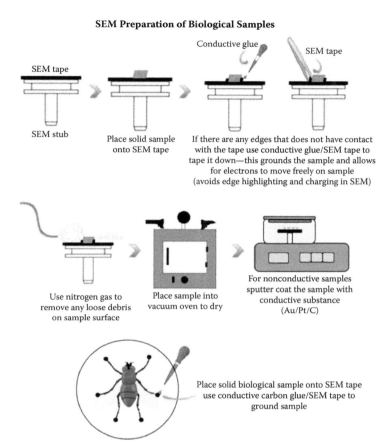

Procedures for handling biological samples on the SEM stub prior to analysis.

5.11.6.2 Biological Samples

Some biological samples (i.e., hair, leaves, insects, fish scales) can be prepared and analyzed without further complicated fixation techniques. Mount the sample on SEM tape and use conductive glue or tape to ensure any edges that do not have contact with the SEM tape are electrically connected. This is important to help ground the sample on the tape and avoid charging and edge highlighting effects. Ensure the biological sample is not loose (use conductive tape to ensure the sample is affixed to the SEM stub) or contain any debris (use nitrogen gas to remove any loose debris from the sample surface). Biological samples are nonconductive and will need to be coated with a conductive material (i.e., Au/Pt or C). Illustrated is a general sample preparation procedure for biological samples.

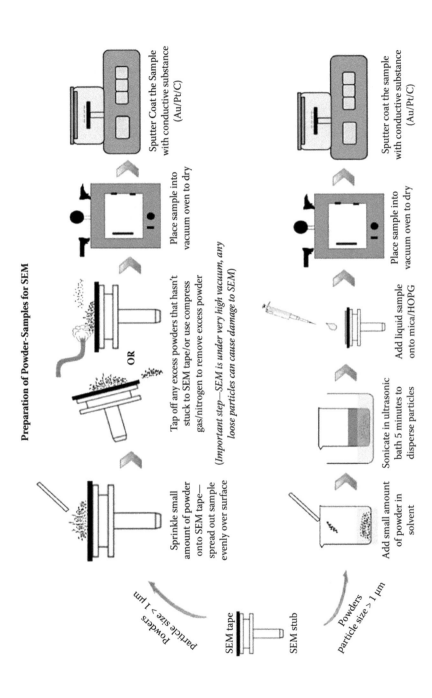

Procedures for handling biological samples on the SEM stub prior to analysis.

5.11.6.3 Powdered Samples

5.11.6.3.1 Powders 1 μm or Larger   For powders that are equal to or greater than **1 μm**, sprinkle a small amount of powder on the SEM tape, making sure the sample is spread evenly over the surface of the SEM stub. Tap off excess powder that has not adhered to the SEM tape or use nitrogen gas to remove the excess powder. This is an important step; the SEM is under high vacuum, and any loose particles can cause damage to the SEM. Place the SEM stub into the vacuum oven to dry and coat the sample with conductive material (i.e., Au/Pt or C). An illustration of a general sample preparation procedure for powdered samples is provided.

Preparation of Liquid Samples for SEM

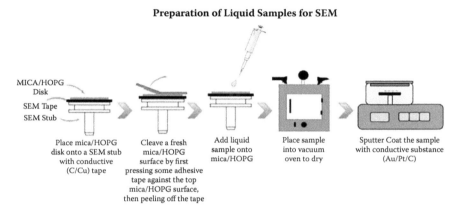

Procedures for handling powder samples on the SEM stub prior to analysis.

5.11.6.3.2   Powders 1 μm or Smaller   For powders that are equal to or less than **1 μm**, add a small amount of powder to a solvent (usually water) and sonicate in an ultrasonic bath for approximately 5 minutes to disperse powder particles. Add liquid powdered sample to an SEM stub that contains freshly cleaved mica/HOPG. Place the SEM stub into the vacuum oven to dry; once the sample is dry, coat it with conductive material (i.e., Au/Pt or C). An illustration of a general sample preparation procedure for powdered samples was indicated in Section 5.11.6.3.2.

5.11.6.4 Liquid Samples

If the sample is a liquid, mount a mica/HOPG disk on the SEM tape and cleave the mica/HOPG disk using double-sided adhesive for a fresh clean surface. Add liquid sample on the SEM stub that contains the mica/HOPG. Place the SEM stub into the vacuum oven to dry; once the sample is dry, coat it with conductive material (i.e., Au/Pt or C). The illustration is of a general sample preparation procedure for liquid samples.

SEM Sample Preparation Record Sheet

Date:

Sample Preparation by:

Sample Description:

Sample Number/Stub Number:

File Name of Sample:

Sample Experimental Conditions:

Sample Preparation Conditions:

Coating: No Coating ☐   Au Coating ☐   Pt Coating ☐   C Coating ☐

Comments

Procedures for handling liquid samples on the SEM stub prior to analysis.

FURTHER READING MATERIAL

1. Voutou, B.; Stefanaki, E.-C.; Giannakopoulos, K. Electron Microscopy: the Basics. *Physics of Advanced Materials Winter School* **2008**, 1–11.
2. Brundle, C. R.; Evans, C. A.; Wilson, S. *Encyclopedia of Materials Characterization: Surfaces, Interfaces, Thin Films*; Gulf Professional: Houston, TX, 1992.
3. Oatley, C. The early history of the scanning electron microscope. *J. Appl. Phys.* **1982,** *53* (2), R1–R13.
4. Danilatos, G. Review and outline of environmental SEM at present. *J. Microsc.* **1991,** *162* (3), 391–402.
5. Suzuki, E. High-resolution scanning electron microscopy of immune-gold labeled cells by the use of thin plasma coating of osmium. *J. Microsc.* **2002,** *208* (3), 153–157.
6. Russell, S. D.; Daghlian, C. P. Scanning electron microscopic observations on deembedded biological tissue sections: comparison of different fixatives and embedding materials. *J. Elec. Microsc. Tech.* **1985,** *2* (5), 489–495.

5.12 SAMPLE ANALYSIS USING ATOMIC FORCE MICROSCOPY

5.12.1 Aim

The aim is to gain hands-on skills and capability in preparing samples for AFM as well as gain valuable experience regarding the workings of an atomic force microscope.

5.12.2 Introduction

Scanning probe microscopes (SPMs) define a broad group of instruments used to image and measure properties of material, chemical, and biological surfaces. SPM images are obtained by scanning a sharp probe across a surface while monitoring and compiling the tip-sample interactions to provide an image. The two primary forms of SPM are STM and AFM. The first STM was developed by IBM Zurich in 1981. Although the ability of the STM to image and measure material surface morphology with atomic resolution has been well documented, only electrical conductors and semiconductors are good candidates for this technique. This significantly limits the materials that can be studied using STM and led to the development, in 1986, of the AFM by Binnig, Quate, and Gerber at Stanford. The sharp probe in this case detects minute forces from the surface under investigation. This enables the detection of atomic-scale features on a wide range of insulating surfaces, including ceramic materials, biological samples, and polymers.

Prior to the invention of SPMs, researchers traditionally used (and still use) a variety of microscopes to image surfaces and measure surface morphology on a microscale. Optical microscopes are the most common instruments available in a general laboratory to image any sample that is not completely optically transparent. Resolution is limited to about 1 μm, and only images and size measurements from features lying in the surface (x-y) plane are obtainable. Furthermore, optical microscopy has a relatively small depth of field. AFM provides a number of advantages over conventional microscopy techniques. AFMs probe the sample and make measurements in three dimensions, x, y, and z (normal to the sample surface), thus enabling the presentation of three-dimensional images of a sample surface. This provides a great advantage over any microscope available previously. The schematic of the AFM investigating a sample is shown in the illustration.

Schematic of Atomic Force Microscopy Operation

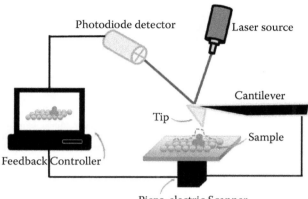

Schematic of the operation of an AFM imaging a sample's surface.

With good samples (clean, with no excessively large surface features), resolution in the x-y plane ranges from 0.1 to 1.0 nm and in the z direction is 0.01 nm (atomic resolution). AFMs do not require a vacuum environment or any special sample preparation, and they can be used in either an ambient or a liquid environment. With these advantages, the AFM has significantly influenced the fields of materials science, chemistry, biology, and physics and the specialized field of semiconductors.

The basic principle of AFM is that a probe is maintained in close contact with the sample surface by a feedback mechanism as it scans over the surface, and the movement of the probe to stay at the same probe-sample distance is taken to be the sample topography. A variety of probes have been used, but the most commonly used are microfabricated silicon (Si) or silicon nitride (Si_3N_4) cantilevers with integrated tips (see the images provided). The bending of the cantilever, normal to the sample surface, is usually monitored by an optical lever consisting of a laser focused on the back of the cantilever, which reflects onto a photodetector (typically a split photodiode). This system magnifies the normal bending of the cantilever greatly and is sensitive to angstrom-level z movements, allowing high-resolution images of the samples at a relatively fast speed depending on the scanning rate.

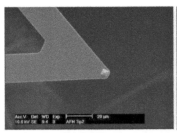 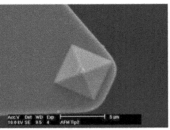

| (a) SEM image of a triangular AFM cantilever consisting of pyramidal silicon nitride tip. Image shows the tip and cantilever used in AFM studies. | (b) SEM image of an AFM silicon nitride tip (pyramidal shaped). Image shows the tip used to scan the surface of a sample in AFM studies at high magnification. |

SEM images of (a) triangular AFM cantilever and (b) AFM silicon nitride tip.

A great advantage of AFM compared to, for example, TEM or SEM is that it is simple to operate in almost any environment, such as not only aqueous solutions but also other solvents, air, vacuum, or other gases. Depending on the details of the experiment, the resolution can be very high. The z (height) resolution is extremely high and can be subangstrom, whereas lateral resolution could be on the order of 1 nm. To obtain such high-resolution images, the state of the sample, the levels of acoustic and electronic noise, and experimental conditions must all be highly optimized. There are generally three main modes of AFM operation, as discussed in Chapter 3: contact, tapping, and noncontact.

In this laboratory session, you will prepare a few samples of NPs from your previous laboratory sessions. These are generally mounted on a mica or HOPG flat surface. In addition, you can also explore natural biological samples such as DNA.

5.12.3 Key Concepts

1. *Operation principles of AFM*

2. *Operation modes of AFM*

3. *Sample preparation procedure for AFM*

5.12.4 Experimental

5.12.4.1 Materials and Equipment

HOPG
Mica
Nitrogen gas

Continued

Milli-Q water
AFM samples to be analyzed

5.12.4.2 Glassware/Equipment

AFM metal stub
AFM stub holder
Scissors
Pasteur pipette
Kimwipes
Spatula
Tweezers
Gloves
Vacuum oven
Double-sided adhesive tape
Ultrasonic bath
Beaker (10 mL)
Digital camera
Micropipettes (10–200 µL)
Micropipette tips (10–200 µL)
Fume hood

5.12.5 Special Safety Precautions

1. Before starting the experiment, look up the Materials Safety Data Sheets for all the chemicals that will be used.

2. Wear gloves, laboratory coats, and safety eyewear at all times while in the laboratory.

3. Use a fume hood if handling fine powders.

5.12.6 Procedure: Sample Analysis Using Atomic Force Microscopy

Prepare AFM samples for analysis with an AFM.

1. Collect the AFM samples required for analysis.

2. Take a photograph of the sample before mounting and after mounting on the AFM stub to include in your laboratory report.

3. Mount double-sided adhesive tape on an AFM metal stub.

The type of sample to be analyzed will determine how it is prepared and mounted to the AFM stub. The following is a general description of how to mount the different types of samples:

4. If analyzing liquid samples, you will need to mount a mica/HOPG disk on the stub, press double-sided adhesive to the top of mica/HOPG surface, then peel off the tape to obtain a clean surface (see the illustration).

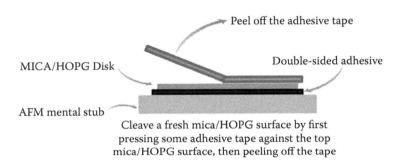

Peel off the adhesive tape

Double-sided adhesive

MICA/HOPG Disk

AFM mental stub

Cleave a fresh mica/HOPG surface by first
pressing some adhesive tape against the top
mica/HOPG surface, then peeling off the tape

Sample preparation procedure for AFM.

5. Use the sample instructions provided in the following for the type of samples to be analyzed.

6. Allow the sample to dry using a vacuum oven.

7. Use nitrogen gas to remove any debris on the sample surface, especially powdered samples.

8. Once a sample is mounted, record the sample in detail using the preparation sheet included.

Types of Samples Prepared during Laboratory Course

Type of Samples to be Analyzed	Samples from the Labs Performed
Powdered samples > 1 μm	ZnO nanorods
Powdered samples < 1 μm	Raw carbon soot
Solid samples	Fingerprint slides, plant leaves slides
Liquid samples	Au NPs, Ag NPs, ZnS NPs, carbon soot, PLGA NPs, bimetallic NPs (Fe@Ag, Fe@Au)
Biological samples	Cells, DNA, and proteins

- The type of sample to be analyzed will determine how it is prepared and mounted.
- **Ensure the sample is flat, stable, and dry and without loose particles.**
- Please record the sample description with as much detail as possible to help with sample analysis later.

5.12.6.1 Liquids

If a sample is a liquid, mount a mica/HOPG disk on double-sided adhesive tape; cleave the mica/HOPG disk by first pressing some adhesive tape against the top of the mica/HOPG surface, then peel off the tape. Add liquid sample on an AFM metal stub that contains freshly cleaved mica/HOPG. Place the AFM stub into a vacuum oven to dry. The figure illustrates a general sample preparation procedure for liquid samples.

AFM Preparation of Liquid Samples

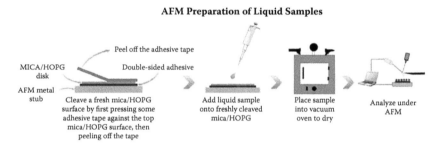

Liquid sample preparation procedure for AFM.

5.12.6.2 Solids

If the sample is solid, mount the sample on the double-sided adhesive tape. Ensure the sample is stable and flat on the adhesive tape. Use nitrogen gas to remove any debris from the sample. The illustration is of a general sample preparation procedure for liquid samples.

AFM Preparation of Solid Samples

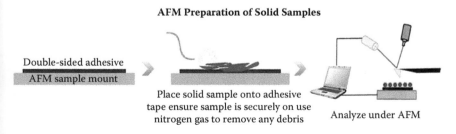

Solid sample preparation procedure for AFM.

5.12.6.3 Powdered Samples

5.12.6.3.1 Powder 1 µm or Larger   Sprinkle a small amount of powder onto the double-sided adhesive tape, making sure the sample is spread evenly over the surface of the stub. Tap off excess powder that has not stuck to the double-sided adhesive tape or use nitrogen to remove excess powder. The illustration provides a general sample preparation procedure for powdered samples.

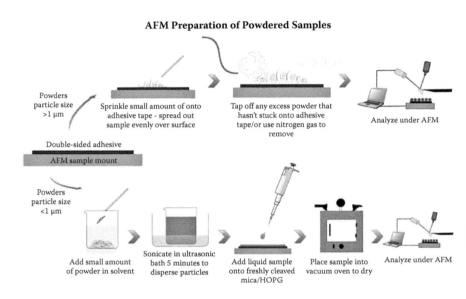

AFM Preparation of Powdered Samples

Powdered sample preparation procedure for AFM.

5.12.6.3.2 Powder 1 µm or Smaller   Add a small amount of powder to a solvent (usually water) and sonicate in an ultrasonic bath for approximately 5 minutes to disperse powdered particles. Add liquid sample to an AFM metal stub that contains freshly cleaved mica/HOPG. Place the AFM stub in a vacuum oven to dry. The powdered sample preparation procedure illustrates pertains to this procedure.

Please record details of how the samples were prepared on a typical AFM sample preparation record sheet as shown in the final illustration for efficient archiving purposes. This will help you greatly for future reference and for writing your scientific report accurately.

AFM Sample Preparation Record Sheet

Date:

Sample preparation by:

Sample description:

File name of sample:

Sample experimental conditions:

Sample preparation conditions:

Comments

A typical AFM sample preparation record sheet.

FURTHER READING MATERIAL

1. Binnig, G.; Rohrer, H.; Gerber, C.; Weibel, E. Surface studies by scanning tunnelling microscopy. *Phys. Rev. Lett.* **1982,** *49* (1), 57–61.
2. Magonov, S. N.; Whangbo, M.-H. *Surface Analysis with STM and AFM: Experimental and Theoretical Aspects of Image Analysis*; Wiley: New York, 2008.
3. Strausser, Y.; Heaton, M. Scanning probe microscopy. *Am. Lab.* **1994,** *26,* 20–29.
4. Magonov, S. N. Surface characterization of msaterials at ambient conditions by scanning tunneling microscopy (STM) and atomic force microscopy (AFM). *Appl. Spectrosc. Rev.* **1993,** *28* (1–2), 1–121.

CHAPTER 6

Nanotechnology and Nanoscience Projects

6.1 INTRODUCTION

Technology can be described as the application of scientific knowledge to the economic growth of manufactured goods, tools, and services. This greatly relies on charting and understanding the vast array of material properties available today and then effectively using these materials accordingly to make the most efficient product or system. Even developing new tools to chart new and exotic properties of nanometer-scale materials for potential future developments still depends on advancements in a number of scientific and engineering fields. To become an independent research scientist, it requires one to have a complex set of skills and capabilities as well as a flair for tackling things that are unknown. It is more than likely that new experimental techniques will need to be developed or existing techniques will need to be adapted. At times, even old proven methods will need to be optimized to tackle new processes. And, in some cases, new instrumentation will need to be designed, built, and fully tested. The proper procedures for using the new instruments and equipment will need to be established so that precise and reliable measurements can be made. This journey is by no means an easy path for any aspiring scientist or engineer because there are inherent uncertainties associated with research, but at the same time, information can be gathered in this scientific expedition and utilized to formulate and test new ideas or hypotheses. A sense of resourcefulness and a set of research skills and abilities are crucial to the development of an independent scientist who will make serious contributions to the advancement of a chosen field of research [1].

So far, you have been involved with performing assigned laboratory experiments (Chapter 5), which were designed to develop your experimental skills and research ability. However, in both academia and industry, a researcher is often required to work on a particular problem with limited supervision, so the researcher needs to develop the ability to think and conduct research independently. In preparation for this type of future career, this chapter presents a number of special projects that incorporate and build from the previous laboratory experiments. This type of approach has provided many undergraduate students with research-based projects that have enhanced their skills and expertise, which has assisted and encouraged them to undertake further postgraduate studies. The nanotechnology-based projects presented in this chapter are intended to provide experience in choosing exciting topics, designing an experiment with the help of literature references, building simple equipment and devices, designing an appropriate experimental procedure, and finally, preparing a proper scientific report as formulated in Chapter 4.

To successfully complete a project, a substantial amount of time should be allocated (15–20 hours of laboratory time at least) so that sufficient information and reliable data can be collected over a typical period of a semester. These projects have been grouped into three main topics, as shown in the Figure 6.1: nanosynthesis, nanocharacterization, and nanoenergy.

The main emphasis of this chapter is in building research skills, independent planning of experiments, acquisition of a set of experimental skills, and good scientific report writing. The essential elements of good scientific writing and its structure have been covered in Chapter 4. Importantly, we have found that some of the research outcomes from the research problems and questions posed in some of the nano projects have resulted in publishable outcomes [2,3].

Some of the projects presented in this chapter can be done in pairs, with both partners working together with either a research assistant and their supervisor or a unit coordinator. All laboratory work should be supervised by an experienced staff member, and the students undertaking the nano projects should be made aware of all the safety procedures and any potential risks or hazards that might be associated with their experimental work

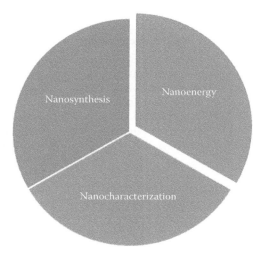

FIGURE 6.1   Main categories of nanotechnology and nanoscience projects.

prior to the commencement of their projects. The students should also be supplied with all Materials Safety Data Sheets of the chemicals they will be using for their projects.

The projects as defined previously are grouped into three main areas: nanosynthesis, nanocharacterization, and nanoenergy. Details are presented in Table 6.1. These areas were selected to give the widest possible range of nanotechnology-based projects for student selection. All projects require hard work, dedication, a sense of purpose, and enthusiasm from both the student and supervisors. In discovery science, there is no certainty or predictability regarding the exact research outcomes that will be achieved from the experimental work; nonetheless, it will be an instructive experience for students to develop their skills and gain experience as a research scientist or engineer.

6.2 NANOSYNTHESIS PROJECTS

In this set of nanosynthesis projects, the main emphasis is to work on synthesizing nanometer-scale materials using various methods that were presented in several of the laboratory sessions. In these projects, the following lines of investigations can be explored and expanded, with related references presented at the end of this chapter and can be used as a starting point for these scientific explorations: gold nanoparticles (Au NPs) [4–7]; silver (Ag) NPs [8–11]; quantum dots (QDs) [12–15];

TABLE 6.1   Nanoprojects and the Related Laboratory Sessions

Nanosynthesis	Nanocharacterization	Nanoenergy
Au NPs (Laboratory 5.1)	FE-SEM of nanophotonic materials (butterfly and beetles) and other nature-based materials (Laboratory 5.11)	Nanocarbon/photothermal applications (Laboratory 5.4)
Ag biosynthesis (Laboratory 5.2)	TLC of leaf extracts (Laboratory 5.4)	DSSC device based on natural dyes and ZnO or QDs (Laboratory 5.3 and Laboratory 5.5)
ZnS/Se QDs Nanoreactors (Laboratory 5.3)	AFM of NPs (Laboratory 5.12)	Microbial fuel cell using nanomaterials or nanocarbon-based electrodes (Laboratory 5.4)
ZnO NPs (Laboratory 5.5)	Raman spectra of CNS or oxidized CNS and graphene (Laboratory 5.10)	
Cu@Au or Cu@Ag bimetallic NPs (Laboratory 5.6)	XRD of NPs (Laboratory 5.11)	
Nanopolymers (Laboratory 5.7)		
Synthesizing alginate/ chitosan nanopolymeric bead NPs (Laboratory 5.9)		
Superhydrophobic surfaces (Laboratory 5.10)		

ZnO NPs [16–19]; bimetallic NPs [20–22]; nanopolymer [23–26]; alginate/chitosan beads [27–29]; and superhydrophobic surfaces [30–33].

6.2.1 Gold NPs (Laboratory 5.1)

1. Creation of Au nanowires using Au seeds from a previous Au NP experiment.

2. Investigation of the strength of selected reducing agents versus a solution of metallic Au salts.

3. Synthesis of Au NPs with the help of a template, such as a nanoporous aluminum oxide membrane.

6.2.2 Silver NPs (Laboratory 5.2)

1. Creation of Ag NPs using Ag NP seeds from a previous experiment.

2. Investigation of the strength of a selected reducing agent versus a solution of metallic Ag salts.

3. Synthesis of Ag NPs with a template, such as a nanoporous aluminum oxide membrane.

4. Investigation of the synthesis of Ag NPs with various plant extracts (indigenous plants can be included as well).

5. Synthesis of Ag NPs with edible material (e.g., seeds, spices, herbs, sugars, etc.).

6.2.3 QDs ZnS/Se NPs (Laboratory 5.3)

1. Synthesis of QDs by the reverse micelle (nano-/microreactors) method with different organic phases and surfactants.

2. Synthesis of QDs with various ratios of Se, S, and Zn and their fluorescent properties.

3. Electrochemical synthesis of Zn/S or Zn/Se QDs as a fast method for generating thin nanomaterial films.

6.2.4 ZnO NPs (Laboratory 5.5)

1. Synthesis of ZnO nanomaterial coatings on substrates such as microscope slides, Si wafers, ITO glass, and so on.

2. Synthesis of ZnO with different basic agents using Zn^{2+} ions.

3. Investigation of the synthesis of ZnO from evaporated Zn-based thin films in air.

6.2.5 Bimetallic Nanoparticles (Laboratory 5.6)

1. Solution synthesis of Cu@Au NPs.

2. Solution synthesis of Cu@Ag, NPs.

3. Investigation of the solution synthesis of magnetic X@Fe NPs, where X denotes a metallic element.

4. Solution synthesis of Fe@X NPs, where X denotes a metallic element.

6.2.6 Nanopolymers (Laboratory 5.7)

1. Solution synthesis of nanopolymers with various monomers, such as polylactide glycolic acid (PLGA), Polyglycolic acid (PGA), poly-2-hydroxyethyl-methacrylate (pHEMA), and others.

2. Synthesis of nanopolymer by other methods (e.g., ultrasound or solvent evaporation).

6.2.7 Alginate/Chitosan Beads (Laboratory 5.9)

1. Synthesis of Alginate/Chitosan beads with model dye molecules as drug delivery models.

2. Study of the kinetics of the drug release profile of X (where X denotes a particular bioactive drug) from alginate beads using the ultraviolet-visible (UV-Vis) technique.

6.2.8 Superhydrophobic Surfaces (Laboratory 5.10)

1. Analysis of natural and synthetic superhydrophobic surfaces.

2. Electrochemical film formation of superhydrophobic surfaces.

3. Contact angle measurements of superhydrophilic and superhydrophobic surfaces.

6.3 NANOCHARACTERIZATION PROJECTS

Nanocharacterization has been and continues to be an important field in the development and application of nanoscience and nanotechnology. Most of the exciting findings in the nanometer world have been made using nanocharacterization tools. In this nanocharacterization section, several nanotechnology projects are proposed. These are based primarily on the direct and indirect imaging techniques discussed in Chapter 3. The specific nanoprojects have been grouped in sections according to the techniques explained in Chapter 3.

6.3.1 Atomic Force Microscopy

The atomic force microscope (AFM) can be used to study and characterize the NPs formed in work delineated in the previous sections. Moreover, the other options available to the instrumentalist in this area, such as spectroscopy and force volume options, can also be used to further evaluate the synthesized NPs and their local properties. Other time-dependent study, such as nanowire formation (and other nanostructures), made with and without DNA can be directly evaluated with the AFM. For more details on the relevant background of the technique and possible applications, please refer to the references at the end of the chapter [34–36].

6.3.2 Scanning Tunneling Microscopy

The scanning tunneling microscope (STM) can also be used to study metallic NPs such as Cu, Ag, and Au metals made by the various synthetic pathways. These can be observed down to atomic resolution, and various options are available with this instrument. For example, local spectroscopy can be used as well when operating an STM to determine other properties of the NPs. Another rich area of study that can be investigated by the STM is that of self-assembly, whereby thiols, for example, can be attached to gold NPs and imaged down to atomic levels. Even thin films with various thiols of different lengths can be studied with the STM on Au, Ag, or Cu thin films. For more details on the relevant background of the technique and possible applications, please refer to the references at the end of the chapter [37–40].

6.3.3 Field Emission Scanning Electron Microscopy

Morphology and size of NPs and nanostructures can be readily evaluated with an advanced field emission scanning electron microscope (FE-SEM). The preparation of a suitable SEM sample is described in detail for several types of samples in Laboratory 5.11, and the procedures can be used in a straightforward manner to study nanomaterials made by various synthetic routes and their experimental parameters evaluated. Further studies with the energy dispersive X-ray system can be used to evaluate other elemental components contributing to the formation of such nanomaterials. Investigation of bionanomaterials such as corals and marine-based calcium carbonates at the nano-/micrometer level can be made by the FE-SEM. Further exploration of nanophotonic materials from insects such as butterflies and beetles can also be investigated with this type of microscopy and the nano-/micrometer features related to the optical properties. For more details on the relevant background of the technique and possible applications, please refer to the references at the end of the chapter [41–43].

6.3.4 Transmission Electron Microscopy

Similar to the FE-SEM, the transmission electron microscope (TEM) can be used to determine the size and morphology of NPs created by various synthetic paths. In addition, the TEM capability of electron diffraction can be used to determine the atomic planes in the NPs imaged. For more details on the relevant background of the technique and possible applications, please refer to the references at the end of the chapter [44–47].

6.3.5 X-ray Diffraction

X-ray diffraction (XRD) is a valuable technique in the arsenal of a material scientist to determine the crystalline phase of NPs and the experimental factors contributing to a particular crystalline phase. This technique can also be used for direct identification of certain materials being synthesized. The various metallic NPs created in the previous stages can be directly evaluated with the XRD technique and the crystalline phase determined. Moreover, the experimental parameters can be related to the creation of a specific phase and accordingly evaluated. For more details on the relevant background of the technique and possible applications, please refer to the references at the end of the chapter [48,49].

6.3.6 Raman Spectroscopy

Raman spectroscopy is an analytical technique commonly used today, and it deals with the vibrational transitions in a molecule or in certain materials. This is especially true for carbon nanomaterials. Raman spectroscopy was used extensively by the earlier pioneers of nanocarbon synthesis and is still in use today for characterizing exciting molecules such as graphene and other important carbon macromolecules or fullerenes. For example, nano- and micrometer-size carbons such as those created in a simple candle flame can be readily evaluated by Raman spectroscopy. Furthermore, these carbons can be modified chemically, oxidized, or functionalized with other molecules. These modified nanocarbons as well as graphene-based materials can then be investigated by Raman spectroscopy to determine the impact of this type of functionalization. Raman spectroscopy can be used to investigate the surface-enhanced Raman scattering (SERS) response of nanomaterials such as Ag, Cu, and Au functionalized with a range of organic molecules. For more details about the technique and some applications, please refer to the references at the end of the chapter [50–52].

6.3.7 UV-Vis Spectroscopy

UV-Vis spectroscopy is a simple technique that can be readily used to evaluate the color solution of Au NPs and those from Ag and Cu NPs.

Experimental parameters such as temperature of synthesis or surfactants/ reducing agent ratio can be readily evaluated with this technique. In addition, dye release profiles from alginate or other polymer beads can be monitored by this technique to provide a time-dye release profile. For more details on the relevant background of the technique and possible applications, please refer to the references at the end of the chapter [53,54].

6.3.8 Thin-Layer Chromatography

The thin-layer chromatographic (TLC) technique is a simple and efficient for evaluating components in a mixture. TLC can be used similar to the manner in Laboratory 5.4 to evaluate the different chlorophyll components present in vegetables such as cabbage leaves of different colors. Even vegetables such as capsicums or carrots of various colors can be tested by the TLC technique. The effects of cooking time and temperature on these vegetables can be investigated, leading to further understanding of human cooking methods. For more details on the relevant background of the technique and possible applications, please refer to the references at the end of the chapter [55–57].

6.4 NANOENERGY PROJECTS

Today, the entire world faces an energy crisis such as never seen before, and many nations have to rely exclusively on costly imported fossil fuels to generate electricity to power their economy and provide energy for the population. Even in advanced economies, fossil fuels are increasingly becoming expensive and have a polluting cost associated with their continued use. Nanotechnology has shown many glimpses of instances for manipulating nanomaterial to create new avenues for sustainable clean energy and potential solutions for the future. In this last section, the nanoenergy project contains a set of projects for students to evaluate the potential of nanotechnology and nanoscience in the area of renewable energy and sustainability. This area is rich in concepts and ideas to be tested in a rigorous and scientific manner, especially by an upcoming scientist. The synthesized nanomaterials in previous laboratory sessions can be further tested for their potential in this area.

For example, the nanocarbons created in Laboratory 5.4 can be used to coat a simple Cu metal foil and the temperature profile plotted versus time when this carbon-coated foil is placed in direct sunlight. The levels of coatings and nature of the nanocarbons contributing to the profile can be investigated. Also, a potential device, similar to a current solar hot water heater, can be made and tested for its power conversion efficiency in sunlight.

Another area of great scientific and research and development interest today is the replication of the photosynthesis process using natural dyes in a modified solar cell. This bio-inspired project is of interest because photosynthesis is an extremely efficient process, which can be seen by the variety of plant and some microbial life on this planet. The dye-sensitized solar cell (DSSC) can be readily created using a combination of nanotitania with various natural dyes, such as strawberry or other natural food/vegetable dyes. These cells can be created and then evaluated for their respective I versus V characteristics and the impact of various experimental parameters investigated along the way.

Fuel cells can be an alternative source of power, and one active area of research and development is in microbial fuel cells (MFCs). In this type of cell, microbes are utilized to generate hydrogen that can be used as a source of fuel. Carbon buckypaper can be readily made from either a commercial source of carbon nanotubes (CNTs) or using graphene-based materials. These can then be characterized with the nanotools described previously and made into electrode substrates for the microbes (either on its own or also doped with metallic NPs). Then, the efficiency of these cells can be evaluated.

REFERENCES

1. Shoemaker, D. P.; Garland, C. W.; Nibler, J. W.; Feigerle, C. S. *Experiments in Physical Chemistry*; McGraw-Hill: New York, 1967.
2. Poinern, G. E.; Brundavanam, R. K.; Mondinos, N.; Jiang, Z.-T. Synthesis and characterisation of nanohydroxyapatite using an ultrasound assisted method. *Ultrason. Sonochem.* **2009**, *16* (4), 469–474.
3. Poinern, G.; Brundavanam, S.; Shah, M.; Laava, I.; Fawcett, D. Photothermal response of CVD synthesized carbon (nano) spheres/aqueous nanofluids for potential application in direct solar absorption collectors: a preliminary investigation. *Nanotechnol. Sci. Appl.* **2012**, *5* (1), 49–59.
4. Gou, L.; Murphy, C. J. Fine-tuning the shape of gold nanorods. *Chem. Mater.* **2005**, *17* (14), 3668–3672.
5. Shankar, S. S.; Rai, A.; Ankamwar, B.; Singh, A.; Ahmad, A.; Sastry, M. Biological synthesis of triangular gold nanoprisms. *Nat. Mater.* **2004**, *3* (7), 482–488.
6. Wu, H.; Ji, X.; Zhao, L.; Yang, S.; Xie, R.; Yang, W. Shape evolution of citrate capped gold nanoparticles in seeding approach. *Colloids Surf., A.* **2012**, *415*, 174–179.
7. Poinern, G. E. J.; Chapman, P.; Le, X.; Fawcett, D. Green biosynthesis of gold nanometre scale plates using the leaf extracts from an indigenous Australian plant *Eucalyptus macrocarpa*. *Gold Bulletin* **2013**, *46* (3), 165–173.

8. Desireddy, A.; Conn, B. E.; Guo, J.; Yoon, B.; Barnett, R. N.; Monahan, B. M.; Kirschbaum, K.; Griffith, W. P.; Whetten, R. L.; Landman, U. Ultrastable silver nanoparticles. *Nature* **2013**, *501*, 399–402.

9. Sharma, V. K.; Yngard, R. A.; Lin, Y. Silver nanoparticles: green synthesis and their antimicrobial activities. *Adv. Colloid Interface Sci.* **2009**, *145* (1), 83–96.

10. Raveendran, P.; Fu, J.; Wallen, S. L. Completely "green" synthesis and stabilization of metal nanoparticles. *J. Am. Chem. Soc.* **2003**, *125* (46), 13940–13941.

11. Poinern, G. E. J.; Chapman, P.; Shah, M.; Fawcett, D. Green biosynthesis of silver nanocubes using the leaf extracts from *Eucalyptus macrocarpa*. *Nano Bulletin* **2013**, *2* (1), 1–7.

12. Steckel, J. S.; Zimmer, J. P.; Coe-Sullivan, S.; Stott, N. E.; Bulović, V.; Bawendi, M. G. Blue luminescence from (CdS) ZnS core–shell nanocrystals. *Angew. Chem. Int. Ed.* **2004**, *43* (16), 2154–2158.

13. Medintz, I. L.; Uyeda, H. T.; Goldman, E. R.; Mattoussi, H. Quantum dot bioconjugates for imaging, labelling and sensing. *Nat. Mater.* **2005**, *4* (6), 435–446.

14. Dabbousi, B.; Rodriguez-Viejo, J.; Mikulec, F. V.; Heine, J.; Mattoussi, H.; Ober, R.; Jensen, K.; Bawendi, M. (CdSe) ZnS core-shell quantum dots: synthesis and characterization of a size series of highly luminescent nanocrystallites. *J. Phys. Chem. B* **1997**, *101* (46), 9463–9475.

15. Kairdolf, B. A.; Smith, A. M.; Stokes, T. D.; Wang, M. D.; Young, A. N.; Nie, S. Semiconductor quantum dots for bioimaging and biodiagnostic applications. *Annu. Rev. Anal. Chem.* **2013**, *6*, 143–162.

16. Tian, Z. R.; Voigt, J. A.; Liu, J.; Mckenzie, B.; McDermott, M. J.; Rodriguez, M. A.; Konishi, H.; Xu, H. Complex and oriented ZnO nanostructures. *Nat. Mater.* **2003**, *2* (12), 821–826.

17. Moezzi, A.; McDonagh, A. M.; Cortie, M. B. Zinc oxide particles: synthesis, properties and applications. *Chem. Eng. J.* **2012**, *185*, 1–22.

18. Hong, R.; Pan, T.; Qian, J.; Li, H. Synthesis and surface modification of ZnO nanoparticles. *Chem. Eng. J.* **2006**, *119* (2), 71–81.

19. Meulenkamp, E. A. Synthesis and growth of ZnO nanoparticles. *J. Phys. Chem. B* **1998**, *102* (29), 5566–5572.

20. Lin, J.; Zhou, W.; Kumbhar, A.; Wiemann, J.; Fang, J.; Carpenter, E.; O'Connor, C. Gold-coated iron (Fe@Au) nanoparticles: synthesis, characterization, and magnetic field-induced self-assembly. *J. Solid State Chem.* **2001**, *159* (1), 26–31.

21. Sra, A. K.; Schaak, R. E. Synthesis of atomically ordered AuCu and AuCu3 nanocrystals from bimetallic nanoparticle precursors. *J. Am. Chem. Soc.*, **2004**, *126* (21), 6667–6672.

22. Toshima, N.; Yonezawa, T. Bimetallic nanoparticles—novel materials for chemical and physical applications. *New J. Chem.* **1998**, *22* (11), 1179–1201.

23. Hu, L.; Iliuk, A.; Galan, J.; Hans, M.; Tao, W. A. Identification of drug targets *in vitro* and in living cells by soluble-nanopolymer-based proteomics. *Angew. Chem. Int. Ed.* **2011**, *50* (18), 4133–4136.

24. Poinern, G. E. J.; Le, X. T.; Shan, S.; Ellis, T.; Fenwick, S.; Edwards, J.; Fawcett, D. Ultrasonic synthetic technique to manufacture a pHEMA nanopolymeric-based vaccine against the H6N2 avian influenza virus: a preliminary investigation. *Int. J. Nanomed.* **2011**, 6, 2167–2174.
25. Balasundaram, G.; Webster, T. J. An overview of nano-polymers for orthopedic applications. *Macromol. Biosci.* **2007**, 7 (5), 635–642.
26. Pinto Reis, C.; Neufeld, R. J.; Ribeiro, A. J.; Veiga, F. Nanoencapsulation I. Methods for preparation of drug-loaded polymeric nanoparticles. *Nanomed. Nanotechnol. Biol. Med.* **2006**, 2 (1), 8–21.
27. Pasparakis, G.; Bouropoulos, N. Swelling studies and *in vitro* release of verapamil from calcium alginate and calcium alginate–chitosan beads. *Int. J. Pharm.* **2006**, 323 (1), 34–42.
28. Mi, F.-L.; Sung, H.-W.; Shyu, S.-S. Drug release from chitosan–alginate complex beads reinforced by a naturally occurring cross-linking agent. *Carbohydrate Polymers* **2002**, 48 (1), 61–72.
29. Xu, Y.; Zhan, C.; Fan, L.; Wang, L.; Zheng, H. Preparation of dual crosslinked alginate–chitosan blend gel beads and *in vitro* controlled release in oral site-specific drug delivery system. *Int. J. Pharm.* **2007**, 336 (2), 329–337.
30. Huovinen, E.; Takkunen, L.; Korpela, T. E.; Suvanto, M.; Pakkanen, T. T.; Pakkanen, T. A. Mechanically robust superhydrophobic polymer surfaces based on protective micropillars. *Langmuir* **2014**, 30, 1435–1443.
31. Poinern, G. E. J.; Le, X. T.; Fawcett, D. Superhydrophobic nature of nano-structures on an indigenous Australian eucalyptus plant and its potential application. *Nanotechnol. Sci. Appl.* **2011**, 4 (1), 113–121.
32. Haghdoost, A.; Pitchumani, R. Fabricating superhydrophobic surfaces via a two-step electrodeposition technique. *Langmuir* **2014**, 30 (14), 4183–4191.
33. Zhang, Y.-L.; Xia, H.; Kim, E.; Sun, H.-B. Recent developments in super-hydrophobic surfaces with unique structural and functional properties. *Soft Matter* **2012**, 8 (44), 11217–11231.
34. Binnig, G.; Quate, C. F.; Gerber, C. Atomic force microscope. *Phy. Rev. Let.* **1986**, 56 (9), 930.
35. Giessibl, F. J. Advances in atomic force microscopy. *Rev. Mod. Phys.* **2003**, 75 (3), 949.
36. Jalili, N.; Laxminarayana, K. A review of atomic force microscopy imaging systems: application to molecular metrology and biological sciences. *Mechatronics* **2004**, 14 (8), 907–945.
37. Stroscio, J. A.; Eigler, D. Atomic and molecular manipulation with the scanning tunneling microscope. *Science* **1991**, 254 (5036), 1319–1326.
38. McNab, I. R.; Polanyi, J. C. Patterned atomic reaction at surfaces. *Chem. Rev.* **2006**, 106 (10), 4321–4354.
39. Edinger, K.; Goelzhaeuser, A.; Demota, K.; Woell, C.; Grunze, M. Formation of self-assembled monolayers of n-alkanethiols on gold: a scanning tunneling microscopy study on the modification of substrate morphology. *Langmuir*, **1993**, 9 (1), 4–8.
40. Tersoff, J.; Hamann, D. Theory and application for the scanning tunneling microscope. *Phy. Rev. Let.* **1983**, 50 (25), 1998–2001.

41. Karan, S.; Mallik, B. Nanoflowers grown from phthalocyanine seeds: organic nanorectifiers. *J. Phys. Chem. C* **2008**, *112* (7), 2436–2447.
42. Bittermann, A. G.; Jacobi, S.; Chi, L. F.; Fuchs, H.; Reichelt, R. Contrast studies on organic monolayers of different molecular packing in FESEM and their correlation with SFM data. *Langmuir* **2001**, *17* (6), 1872–1877.
43. Sharma, S.; Rasool, H. I.; Palanisamy, V.; Mathisen, C.; Schmidt, M.; Wong D. T.; Gimzewski, J. K. Structural-mechanical characterization of nanoparticle exosomes in human saliva, using correlative AFM, FESEM, and force spectroscopy. *ACS Nano.* **2010**, *4* (4), 1921–1926.
44. Williams, D. B.; Carter, C. B. *The Transmission Electron Microscope*; Springer: New York, 1996.
45. Wang, Z. L. Transmission electron microscopy of shape-controlled nanocrystals and their assemblies. *J. Phys. Chem B.* **2000**, *104*, 1153–1175.
46. Pietra, F.; Rabouw, F. T.; Evers, W. H.; Byelov, D. V.; Petukhov, A. V.; de Mello Donegá, C.; Vanmaekelbergh, D. l. Semiconductor nanorod self-assembly at the liquid/air interface studied by *in situ* GISAXS and *ex situ* TEM. *Nano Lett.* **2012**, *12* (11), 5515–5523.
47. Lee, J.; Saha, A.; Pancera, S. M.; Kempter, A.; Rieger, J.; Bose, A.; Tripathi, A. Shear free and blotless cryo-TEM imaging: a new method for probing early evolution of nanostructures. *Langmuir* **2012**, *28* (9), 4043–4046.
48. Zhu, X.; Birringer, R.; Herr, U.; Gleiter, H. X-ray diffraction studies of the structure of nanometer-sized crystalline materials. *Phys. Rev. Lett. B* **1987**, *35* (17), 9085.
49. Kahle, M.; Kleber, M.; Jahn, R. Review of XRD-based quantitative analyses of clay minerals in soils: the suitability of mineral intensity factors. *Geoderma* **2002**, *109* (3), 191–205.
50. Dresselhaus, M. S.; Jorio, A.; Hofmann, M.; Dresselhaus, G.; Saito, R. Perspectives on carbon nanotubes and graphene Raman spectroscopy. *Nano Lett.* **2010**, *10* (3), 751–758.
51. Kneipp, K.; Kneipp, H.; Itzkan, I.; Dasari, R. R.; Feld, M. S. Ultrasensitive chemical analysis by Raman spectroscopy. *Chem. Rev.* **1999**, *99* (10), 2957–2976.
52. Haynes, C. L.; McFarland, A. D.; Duyne, R. P. V. Surface-enhanced Raman spectroscopy. *Anal. Chem.* **2005**, *77* (17), 338A–346A.
53. Jana, N. R.; Gearheart, L.; Murphy, C. J. Wet chemical synthesis of silver nanorods and nanowires of controllable aspect ratio. *Chem. Commun.* **2001**, (7), 617–618.
54. Haiss, W.; Thanh, N. T.; Aveyard, J.; Fernig, D. G. Determination of size and concentration of gold nanoparticles from UV-Vis spectra. *Anal. Chem.* **2007**, *79* (11), 4215–4221.
55. Bolliger, H.; Brenner, M.; Gänshirt, H. *Thin-Layer Chromatography*; ACS: Washington, DC, 1965.
56. Anwar, M. Separation of plant pigments by thin layer chromatography. *J. Chem. Educ.* **1963**, *40* (1), 29.
57. Jeffrey, S. W. Quantitative thin-layer chromatography of chlorophylls and carotenoids from marine algae. *Biochim. Biophys. Acta Bioenerg.* **1968**, *162* (2), 271–285.

Index

A

Abstract, 81
Acid release, alginate capsule laboratory, 167–171
Acknowledgment section, 85
Adhesive proteins, 9–10
Agar solution, 163
Alarm signal, 78
Alfalfa, 98
Alginate (ALG) capsules
  introductory information, 156–157
  laboratory
    acid release studies, 167–171
    aim, 155
    dye release studies, 165–167
    formation and encapsulation, 164–165
    glassware/equipment, 158–159
    key concepts, 158
    materials/reagents, 158
    preparation of solutions and ALG beads, 160–164
    procedure, 159
    special safety precautions, 159
  nanosynthesis projects, 212
Aluminum oxide (Al_2O_3, alumina), 23
Anodic aluminum oxide templates, 37
Antimicrobial activity, silver, 97
Anti-Stokes-Raman scattering, 60
Atomic force microscopy (AFM), 16, 41, 42t, 49–53, 198–200
  bimetallic NPs, 139
  fingerprint analysis, 153
  gold NP characterization, 95
  polymer NP characterization, 146
  projects, 212–213

sample analysis laboratory
  aim, 198
  introductory information, 198–200
  key concepts, 200
  liquid samples, 203
  materials and equipment, 200–201
  powdered samples, 204
  procedure, 201–203
  record sheet, 205
  solid samples, 203
  special safety precautions, 201
silver NP characterization, 104
zinc oxide nanorod characterization, 132
zinc sulfide NP characterization, 115
Atomic-scale matter manipulation, 16, 47

B

Backscattered electrons (BSEs), 45
Bakelite, 25
Band gap, 106
Beer Lambert's law, 57
Bimetallic nanoparticles, laboratory synthesis, 133–139
  aim, 133
  characterization methods, 139
  glassware/equipment, 134–135
  introductory information, 133–134
  key concepts, 134
  materials/reagents, 134
  procedure, 135
    Fe@Ag NPs, 138–139
    Fe@Au NPs, 136–137
  projects, 211
  special safety precautions, 135

Biodegradable drug delivery systems, 26, 140, 155–156, *See also* Alginate (ALG) capsules
Biological mimicry, 6
Biological sample analysis, SEM, 193–194
Bird feathers, 6
Borohydride, 89, 134
Bottom-up synthesis approaches, 32
  colloids, 34–35
  laser ablation, 36
  microwave-based, 39, 124–125
  sol-gel, 35
  spray pyrolysis, 38
  sputtering, 36
  template-based, 37
  ultrasound-based, 38–39
  vapor deposition, 36
Bragg equation, 54
Bragg reflection, 53
Brownian motion, 62, 107
Buckminsterfullerene (C_{60} or buckyballs), 18, 28, 116

C

Cadmium selenium (CdSe) quantum dots, 26, 27, 38, 149
Cadmium sulfide (CdS) quantum dots, 38, 149
Calcium chloride solution, 163
Candle soot, carbon nanoparticle synthesis, 117–120
Capping agents, 34, 88
Carbon nanoparticles (CNPs) or nanomaterials, 18–19, 27–31
  characterization methods, 122
  laboratory synthesis
    aim, 116
    glassware/equipment, 118
    introductory information, 116–117
    key concepts, 117
    materials/reagents, 117
    procedure, 118
    separation using TLC, 121–122
    special safety precautions, 118
    synthesis from candle soot, 118–120
  synthesis approaches, 116–117

Carbon nanotubes (CNTs), 18, 28–30
  microbial fuel cells and, 216
  synthesis techniques, 37
Catalysis
  bimetallic NPs and, 133
  copper properties, 22
  gold activity, 21
  surface area and, 17
Catenation, 27
Cell culture substrate, 23
Celllulose, 8
Cellulose, 25
Characterization methods, 15–16, *See also specific methods*
  carbon NPs, 122
  dynamic light scattering, 61–63
  Fe@Ag and Fe@Au bimetallic NPs, 139
  fingerprint analysis, 153
  gold NPs, 95
  laser light and the Tyndall effect, 89
  microscopy, 15–16, 41–53, *See also* Microscopy techniques
  nanocharacterization projects, 212–215
  Raman spectroscopy, 60–61
  sample analyses laboratory, 187–206
  silver NPs, 104
  thin-layer chromatography, 58–60
  UV-Vis spectroscopy, 57–58
  X-ray diffraction, 53–56
  zinc sulfide NPs, 115
Chemical hazards, 70–71
  Materials Safety Data Sheet (MSDS), 68, 73
  pollutant detoxification, 133
Chemical vapor deposition (CVD), 36, 107
Chitosan (CS), 156
  citric acid solution, 163
Chlorauric acid ($HAuCl_4$), 34, 88, 89
Chlorinated hydrocarbons, 133
Chromatography, 58, *See also* Thin-layer chromatography
Coffee extract, 99
Collagen, 10
Colloidal particle light scattering, 89
Colloid-based synthesis techniques, 34–35
Color properties, 21, 88, 95, 104, 115, 132

Composite materials, 19
Conclusion section, 83
Confocal microscopy, 42t
Contact angle and hydrophobicity, 173
   laboratory procedures, 179–182, 186
Copper (Cu), 19
   bimetallic nanoparticle synthesis
       projects, 211
   nanoparticles, 22
Cyanoacrylate, 149

D

Detoxification applications, 22
Diabetes treatment applications, 8
Diamond, 28
Discussion section, 83
DOPA, 10
Drug delivery systems
   alginate capsule synthesis and drug/
       dye release profiles, 155–172,
       See also Alginate (ALG)
       capsules
   Au NPs, 20
   biodegradable systems, 140, 155–156
   insulin pump, 8
   polymers, 26, 140, 155
Dye release, alginate capsule laboratory,
       165–167
Dye-sensitized solar cells (DSSCs),
       24, 216
Dynamic light scattering (DLS), 42,
       61–63, 72

E

Electrochemical fabrication, 107
Electron flow, 6
Electron microscopy, *See* Scanning
       electron microscopy
Energy-related nanoscience projects,
       215–216
Epitaxy, 34
Eucalyptus leaf, 11–12, 174–175
Evacuation procedures, 78
Eye protection, 69–70, 76
Eyewash station, 70

F

Faraday, Michael, 34, 88, 97
Feathers, 6
Ferric oxide (Fe_2O_3), 24
Ferrofluids, 25
Feynman, Richard, 2
Field emission scanning electron
       microscopy (FE-SEM), 42t, 43
   carbon NP characterization, 122
   Fe@Ag and Fe@Au bimetallic NPs, 139
   fingerprint analysis, 153
   gold NP characterization, 95
   polymer NP characterization, 146
   projects, 213
   silver NP characterization, 104
   zinc oxide nanorod characterization, 132
   zinc sulfide NP characterization, 115
Fingerprint analysis, 148–149
   characterization methods, 153
   introductory information, 147–148
   laboratory
      aim, 147
      carbon toner fingerprint
         preparation, 151–152
      glassware/equipment, 149–150
      key concepts, 149
      materials/reagents, 149
      slide preparation, 150–151
      special safety precautions, 150
      superglue fuming process, 152
Fire hazards, 72–73
   labeling system, 73–74
Flame pyrolysis, 38
Fluorescent carbon nanoparticles,
       laboratory synthesis, 116–123,
       See also Carbon nanoparticles
       (CNPs) or nanomaterials,
       laboratory synthesis
Fluorescent microscopy, 42t
Fluorescent nanoparticles, 116
   quantum dots, 27
Footwear, 76
Forensic applications, *See* Fingerprint
       analysis
Fuel cells, 216
Fullerenes, 18, 28
Fume cupboard, 76

G

Gas chromatography (GC), 58
Gelatin, 156
    solution preparation, 163
Geranium, 98
Glassware hazards, 71–72
Gold (Au) nanomaterials, 19–21, 88
    Au properties/uses, 19–20, 87–88
    bimetallic Fe/Au NP laboratory
        synthesis, 133–139, *See also*
        Bimetallic nanoparticles,
        laboratory synthesis
    characterization methods, 95
    colloid-based synthesis technique,
        34–35
    color, 21, 88, 95
    general uses, 88
    laboratory synthesis
        aims, 87
        glassware/equipment, 90
        key concepts, 89–90
        materials/reagents, 90
        procedure, 91–95
        special safety precautions, 90–91
    nanosynthesis projects, 210
    red phosphorus-based synthesis
        method, 98
    synthesis approaches, 88
    zeta potential, 63
Gold nanowires, 6
Good laboratory practices, 66
Graphene, 30–31, 116
    carbon nanohorns, 19
    microbial fuel cells and, 216
Graphite, 27
Grätzel cells, 24
Green chemistry, 35, 97–99
Green tea, Ag NP biosynthesis method,
    103–104
Guar gum, 156
    solution preparation, 163
Gutta percha, 25

H

Harvard system of citation, 84
Hazard labeling system, 73–74

Hazard signs, 67–68*f*
High pressure liquid chromatography
    (HPLC), 58
Humbolt squid beak, 7–8
Hydrophilic surfaces, 173, 181–182
    laboratory
        contact angle estimation, 179–181
        leaf preparation, 177–178
        leaf wax self-replicating properties,
            180–181
        self-cleaning experiment, 182–185
Hydrophobicity, 173, *See also*
    Superhydrophobic surfaces

I

Indian watercress, 174
Insulin, 98
Insulin pump, 8
Integrated circuit chip fabrication, 33
Introduction of report, 81–82
Iron (Fe) nanoparticles, 22–23, 139
    ultrasonic synthesis, 38
Iron/gold and iron/silver bimetallic
        nanoparticles, laboratory
        synthesis, 133–139, *See also*
        Bimetallic nanoparticles,
        laboratory synthesis
Iron oxides, 24–25
Iron/palladium bimetallic
        nanoparticles, 133

K

Kohlrabi, 11

L

Lab-on-a-chip (LOC) devices, 33
Laboratory notebook, 79, 80
Laboratory safety, 65–70
    chemical hazards, 70–71
    evacuation procedures, 78
    eye protection, 69–70, 76
    fire hazards, 72–73
    glassware hazards, 71–72
    Materials Safety Data Sheet (MSDS),
        68, 73

NFPA hazard labeling system, 73–74
protective clothing, 69
safety and hazard signs, 67–68f
specific laboratory precautions
    alginate beads, 159
    atomic force microscopy, 201
    bimetallic NP synthesis, 135
    carbon NP synthesis, 118
    fingerprint analysis, 150
    gold NP synthesis, 90–91
    polymeric NP synthesis, 144
    scanning electron microscopy, 190
    silver NP synthesis, 100
    superhydrophobic leaf surfaces
        investigation, 176
    zinc oxide nanorod synthesis, 126
    zinc sulfide NP synthesis, 109
    summary of important rules, 75–76
    in teaching laboratories, 76–78
Laser ablation, 36, 107
Laser light, 72
    Tyndall effect, 89
Leaf preparation procedure,
        superhydrophobic surfaces lab,
        177–179
Light-emitting diodes, 124
Liquid sample analysis
    AFM, 203
    SEM, 195–196
Locard's principle, 147
Lotus leaf and lotus effect, 10–11, 173–174,
        See also Superhydrophobic
        surfaces
Low-pressure chemical vapor deposition
        (LPCVD), 36
Lycargus Cup, 19–20

M

Magnetite (Fe$_2$O$_4$), 24, 38
Materials and methods section, 82
Materials Safety Data Sheet (MSDS), 68,
        73, 209
Medical applications, See also Drug
        delivery systems
    biological adhesives, 10
    nano aluminum oxide, 23
    nanocopper, 22

    nanoiron, 22–23
    nanosilver, 21, 97
    polymers, 26
Melting point, 20
Metal nanomaterials, 19–23, See also
        specific types
Metal organic chemical vapor deposition
        (MOCVD), 36
Metal oxide nanomaterials, 23–25,
        See also specific types
Micelles, 32
Microbial fuel cells (MFCs), 216
Microelectromechanical systems
        (MEMSs), 33
Microscopy techniques, 15–16, 41–53,
        See also Atomic force microscopy;
        Optical microscopy; Scanning
        electron microscopy; Scanning
        tunneling microscopy;
        Transmission electron
        microscopy
    atomic-scale matter manipulation, 16, 47
    chart of, 42t
    nanocharacterization projects, 212–214
    sample analyses laboratory, 187–206
Microwave-based synthesis, 39, 124–125
    ZnO nanorods, 129–130
Mie scattering, 62
Milling, 32
Mimicry, 6
Molecular beam epitaxy (MBE), 34
Mottlecah, 11–12, 174–175
Multiwall carbon nanotubes (MWNTs), 29
Mussel adhesive proteins (MAPs), 9–10
Mussel beard, 8–10

N

Nanoelectromechanical systems
        (NEMSs), 33
Nanoenergy projects, 215–216
Nanomaterial synthesis, See Synthesis
        techniques
Nanometer-scale materials, 5–6, 15, 16,
        See also specific nanomaterials
    characterization methods, See
        Characterization methods
    composite materials, 19

metal oxides, 23–25
metals, 19–23
nanocarbons, 27–31, *See also* Carbon
    nanoparticles (CNPs) or
    nanomaterials
natural materials, *See* Natural
    nanomaterials
polymers, 25–26, *See also* Polymeric
    materials
quantum dots, 26–27
size and surface area relationships,
    16–18
synthesis, *See* Synthesis techniques
types, 17–18
Nanorods
    gold, 20
    characterization methods, 132
    synthesis methods, 23, 37, 38
    ZnO lab, 124–132
Nanoscience, 4–5
Nanotechnology, 2–5
    background literature, 3–4t
    definition, 2, 4
Nanotechnology and nanoscience
        projects, 207–209
    nanocharacterization, 212–215
    nanoenergy, 215–216
    nanosynthesis, 209–212
Nanotubes, 18, 23, 28–30, 37, 47, 116
Nanowires, 6, 23, 37, 132, 213
National Aeronautics and Space
        Administration (NASA), 24–25
National Fire Protection Association
        (NFPA) hazard labeling system,
        73–74
Natural nanomaterials, 6–7
    Humbolt squid beak, 7–8
    mussel beard, 8–10
    superhydrophobic leaf surfaces, 10–12
Neem, 98
Nobel Prize awards, 31, 116

O

Optical microscopy, 42t, 198
    AFM advantages versus, 198
    fingerprint analysis, 153
    SEM advantages versus, 188

Optical properties
    gold and plasmon resonance, 21
    quantum dots, 106
    silver photochemistry, 21–22
    UV-blocking, 24
Organic dyes, 27

P

Particle size and band gap relationship, 106
Particle size and surface area
        relationships, 16–18
Pen lasers, 72
Phenolphthalein solution, 163
Photochromic glass, 22
Photography applications, 21–22
Photolithography, 32, 33
Photoluminescent nanomaterials, 116, 149
Photon correlation spectroscopy (PCS), 62
Photosynthesis, 216
Physical vapor deposition (PVD), 36
Plagiarism, 83–84
Plasma-enhanced chemical vapor
        deposition (PECVD), 36
Plasmon resonance, 21, 61
Poly-2-hydroxyethyl-methacrylate
        (pHEMA), 26, 156
    solution preparation, 163
Poly (lactide-co-glycolide) acid (PLGA), 26
    characterization methods, 146
    laboratory synthesis, 140–146
        aim, 140
        glassware/equipment, 143
        introductory information, 140–141
        key concepts, 143
        materials/reagents, 143
        NPs preparation, 145–146
        solution preparation, 144–145
        special safety precautions, 144
        spontaneous emulsification solvent
            diffusion (SESD) method,
            141–142
        surfactant solution preparation, 145
Polymeric materials, 25–26
    biocompatibility/biodegradability, 26,
        140, 156
    biological adhesives, 10
    characterization methods, 146

nanosynthesis projects, 212
PLGA laboratory synthesis, 140–146,
    See also Poly(lactide-
    co-glycolide) acid
properties/uses, 140
spontaneous emulsification solvent
    diffusion (SESD) method, 141–142
Polymethyl methacrylate (PMMA), 26
Polystyrene (PS), 26
Polyvinyl alcohol (PVA) solution, 145
Polyvinyl chloride (PVC), 26
Pore size distribution, 37
Powdered sample analysis
    AFM, 204
    SEM, 195
Powder X-ray diffraction, 53
Projects, *See* Nanotechnology and
    nanoscience projects
Protective clothing, 69

Q

Quantum confinement, 106
Quantum dots (QDs), 26–27, 106
    fingerprint visualization
        applications, 149
    molecular beam epitaxy, 34
    nanosynthesis projects, 211
    optical properties, 106
    synthesis approaches, 107, *See also* Zinc
        sulfide (ZnS) nanoparticles,
        laboratory synthesis
    ultrasonic synthesis, 38
Quantum tunneling, 47
Quantum well devices, 34
Quasi-elastic light scattering (QELS), 62

R

Raman spectroscopy, 42, 60–61
    projects, 214
    use of lasers, 72
Rayleigh scattering, 60
Red phosphorus, 98
References, 83–84
Report writing, 78–85
Research scientists, characteristics and
    abilities, 207–208

Results section, 82
Reverse micellar synthesis approach, 32,
    107–108
    laboratory procedure, 110–114
Rubber, 25

S

Safety, *See* Laboratory safety
Safety and hazard signs, 67–68*f*
Scanning electron microscopy (SEM), 41,
    43–46, 187–188, *See also* Field
    emission scanning electron
    microscopy (FE-SEM)
    projects, 213
    sample analysis laboratory, 187–197
        aim, 187
        biological samples, 193–194
        glassware/equipment, 189
        key concepts, 189
        liquid samples, 195–196
        materials/reagents, 189
        powdered samples, 195
        procedure, 190–192
        solid samples, 192
        special safety precautions, 190
Scanning probe microscopes (SPMs),
    16, 198
Scanning tunneling microscopy (STM),
    16, 41, 42*t*, 47–49, 198
    gold nanoparticle characterization, 95
    projects, 213
    silver nanoparticle characterization, 104
Scherrer's formula, 56
Scientific report writing, 78–85
Scratch-resistant surfaces, 23
Sea cucumber, 8
Secondary electrons (SEs), 45
Self-cleaning surfaces, 11, 173–175, *See also*
    Superhydrophobic surfaces
    laboratory
        aim, 173
        experiment, 182–185
        introductory information, 173–175
    nanometer-scale titania, 24
Semiconductor materials, 106–107
    carbon nanotubes, 30
    fullerenes, 28

nanometer-scale titania, 24
particle size and bandgap
relationship, 106
quantum dots, 26–27, 106,
See also Zinc sulfide (ZnS)
nanoparticles, laboratory
synthesis
Silica (SiO$_2$), 149
Silver (Ag) nanoparticles, 21–22
Ag characteristics/uses, 19, 21–22, 95
bimetallic Fe/Ag NP laboratory
synthesis, 133–139, *See also*
Bimetallic nanoparticles,
laboratory synthesis
characterization methods, 104
laboratory synthesis
aim, 97
glassware/equipment, 99
green chemistry principles, 97–99
key concepts, 99
materials/reagents, 99
procedure, 100
safety precautions, 100
using green tea, 103–104
using plant/leaf extracts, 101–102
nanosynthesis projects, 211
Single-wall carbon nanotubes (SWNTs),
29–30
Size scales, 5*t*
Sodium borohydride, 89, 134
Sodium citrate, 34–35
Sodium hydroxide solution, 164
Solar cells, 24, 27, 216
Sol-gel materials, 35
Solid sample analysis
AFM, 203
SEM, 192
Solvothermal methods, 107
Spectroscopy, 57, *See also* Raman
spectroscopy; Ultraviolet-visible
(UV-Vis) spectroscopy
photon correlation, 62
Spontaneous emulsification solvent
diffusion (SESD) method,
141–142
laboratory procedure, 144–146
Spray pyrolysis, 38
Sputtering, 36

Stokes-Einstein equation, 62
Stokes-Raman scattering, 60
Sunscreens, 24
Superglue method, fingerprint analysis,
148–149, 152
Superhydrophobic surfaces, 173–175
eucalyptus leaf, 11–12
laboratory
aim, 173
contact angle estimation, 179–182
glassware/equipment, 176
introductory information, 173–175
key concepts, 175
leaf preparation, 177–179
leaf wax self-replicating properties,
180–182
materials/reagents, 175–176
self-cleaning experiment, 182–185
special safety precautions, 176
lotus leaf/lotus effect, 10–11, 173–174
nanosynthesis projects, 212
self-cleaning properties, 11
Surface area and size relationships,
16–18
Surface plasmon resonance (SPR), 21, 61
Surface profilometer, 47
Surfactants, 34
polyvinyl alcohol solution
preparation, 145
Sutures, 26
Synthesis techniques, 19, 31–32, *See also*
specific nanomaterials
bottom-up approaches, 32
colloids, 34–35
laser ablation, 36
microwave-based, 39, 124–125
sol-gel, 35
spray pyrolysis, 38
sputtering, 36
template-based, 37
ultrasound-based, 38–39
vapor deposition, 36
green chemical approach, 35
microwave-based, 124–125
top-down approaches, 32
milling, 32
molecular beam epitaxy, 34
photolithography, 33

Synthesis techniques, laboratories,
See also specific nanomaterials
  bimetallic NPs, 133–139
  drug-/dye-loaded alginate capsules,
    155–172
  fluorescent carbon NPs, 116–123
  gold NPs, 87–96
  polymeric NPs, 140–146
  projects, 209–212
  silver NPs, 97–105
  zinc oxide nanorods, 124–132
  zinc sulfide NPs, 106–115

T

Taniguchi, Norio, 2
Taro, 11, 174
Tea leaves, source of nanoparticles, 98–99,
    103–104
Technology, 207
Template-based synthesis techniques, 37
Thin-layer chromatography (TLC), 42
  carbon nanoparticle synthesis
    approach, 121–122
  projects, 215
Tissue engineering, 23
Titanium dioxide (TiO_2, titania), 24, 216
Top-down synthesis approaches, 32
  milling, 32
  molecular beam epitaxy, 34
  photolithography, 33
Transmission electron microscopy (TEM),
    41, 42t, 46–47
  carbon NP characterization, 122
  Fe@Ag and Fe@Au bimetallic NPs, 139
  gold NP characterization, 95
  polymer NP characterization, 146
  projects, 213–214
  silver NP characterization, 104
  zinc oxide nanorod characterization, 132
  zinc sulfide NP characterization, 115
Trichloroethane (TCE), 133
Tyndall effect, 89

U

Ultrasonic synthesis, 38–39
Ultraviolet (UV) light blocking, 24

Ultraviolet-visible (UV-Vis) spectroscopy,
    42, 57–58, 95, 115, 122, 132,
    214–215

V

Vapor deposition, 36
Vapor liquid solid (VLS) process, 37
Volume heating, 124–125
Vulcanization, 25

W

Waxes, superhydrophobic properties,
    11–12, 174, *See also*
    Superhydrophobic surfaces
  self-replicating properties and contact-
    angle estimation, 181–182

X

X-ray crystallography, 53
X-ray diffraction (XRD), 42, 53–56
  projects, 214

Z

Zeta potential, 63
Zinc acetate dihydrate, 35
Zinc oxide (ZnO), 24, 35, 124–125
  characterization methods, 132
  nanorod synthesis laboratory
    aim, 124
    glassware/equipment, 125–126
    introductory information,
      124–125
    key concepts, 125
    materials/reagents, 125
    microwave method, 129–130
    observation by UV and optical
      microscopy, 131
    procedure, 126–131
    special safety precautions, 126
    ZnO solution preparation,
      127–128
  nanosynthesis projects, 211
Zinc sulfide (ZnS), fingerprint
    visualization application, 149

Zinc sulfide (ZnS) nanoparticles,
    laboratory synthesis
  aim, 106
  glassware/equipment, 109
  introductory information, 106–108
  materials/reagents, 109
  nanocrystal characterization, 115
  procedure, 110–114
  reverse micellar approach, 107–108
  special safety precautions, 109
  synthesis approaches, 107
Zinc sulfide (ZnS) quantum dots, 26,
    106, 211

For Product Safety Concerns and Information please contact our EU
representative GPSR@taylorandfrancis.com
Taylor & Francis Verlag GmbH, Kaufingerstraße 24, 80331 München, Germany